ENVIRONMENT, ETHICS AND THE CORPORATION

Environment, Ethics and the Corporation

Grant Ledgerwood

and

Arlene Idol Broadhurst

First published in Great Britain 2000 by
MACMILLAN PRESS LTD
Houndmills, Basingstoke, Hampshire RG21 6XS and London
Companies and representatives throughout the world

A catalogue record for this book is available from the British Library.

ISBN 0–333–68526–1

First published in the United States of America 2000 by
ST. MARTIN'S PRESS, INC.,
Scholarly and Reference Division,
175 Fifth Avenue, New York, N.Y. 10010

ISBN 0–312–23010–9

Library of Congress Cataloging-in-Publication Data
Ledgerwood, Grant, 1941–
Environment, ethics and the corporation / Grant Ledgerwood and Arlene Idol Broadhurst.
p. cm.
Includes bibliographical references and index.
ISBN 0–312–23010–9 (cloth)
1. Industrial management—Environmental aspects. 2. Social responsibility of business. I. Broadhurst, Arlene Idol. II. Title.

HD30.255 .L428 1999
658.4'08—dc21

99–049280

This book is printed on paper suitable for recycling and made from fully managed and sustained forest sources.

10 9 8 7 6 5 4 3 2 1
09 08 07 06 05 04 03 02 01 00

Printed and bound in Great Britain by
Antony Rowe Ltd, Chippenham, Wiltshire

Tom Peters, some time in the 1990s:

'We are re-engineering the planet.'

Contents

List of Tables, Figures and Boxes

Tables

Figures

Boxes

List of Cases

Preface and Acknowledgements

The range of topics and references in this book indicate the width of canvas we have attempted to sketch. Interrelating the worlds of international relations (AB) and strategic business planning (GL), the planetary development process is full of both hope and threats. In this book we emphasise hope.

We are grateful to the initial commissioning editor, Jane Powell, and her successor, Samantha Whittaker, for allowing us time to forge the initial project. Colleagues at Greenleaf Publishing Sheffield have proven a source of ideas and stimulus, especially John Stuart, now managing director and editor of Greenleaf. His leadership, assistance and support for the journal *Greener Management International Journal of Corporate Environmental Strategy and Practice* (*GMI Journal*), for which GL has been general editor in recent years, has been exemplary. We cite in this text numerous examples of research originally appearing in GMI Journal.

At the University of Greenwich Business School, Sue Millar, head of school, has been an open advocate and strong supporter of the research focus from which this book emanates. She has been unfailingly committed to our research and we appreciate her dedication. We also gratefully acknowledge support of staff at the University of Greenwich Kings Hill Centre, in particular that of Jane Cunningham and Michael Longman. On behalf of the publisher, Keith Povey provided valuable criticism and editing of our final text; we are grateful for his careful scrutiny.

Grant Ledgerwood acknowledges the irreplaceable support of his wife and colleague, Elizabeth Street, in this as in so many other endeavours over twenty-five years.

Arlene Broadhurst would like to thank her husband, John Byrne Broadhurst, for providing technical advice, assistance and unflagging encouragement. Grant Ledgerwood, Elizabeth Street and their children, Kate and Matt, offered hospitality and assistance to a new arrival in Kent.

In spite of all this help, surviving errors remain the authors' responsibility.

GRANT LEDGERWOOD
ARLENE IDOL BROADHURST

An earlier version of Chapter 4 first appeared as Arlene I. Broadhurst and G. Ledgerwood, 'Environmental Diplomacy of States, Corporations, and Non-Governmental Organizations: The World-Wide Web of Influence', *International Relations*, vol. XIV, no. 2 (August 1998). We are grateful to the David Davies Memorial Institute of International Studies, London, for permission to use this material. The authors and the publishers are grateful to Elsevier Science for permission to use material which appeared in an earlier form in the journal *Cities*, vol. 11(3), G. Ledgerwood, I. Haywood, N. Sheppard and E. Street, 'Environmental Assessment as a Real Estate Management Protocol', pp. 181–94, copyright © 1994, with permission from Elsevier Science Ltd, The Boulevard, Langford, Lane, Kidlington OX5 1GB, UK.

Introduction

Environment, Ethics and the Corporation takes a particular view of the corporate environmental revolution which is underway in the closing years of the twentieth century.[1] The 1,000 largest corporations in the world drive international investment. Thereby, these businesses have a more direct impact on planetary environment than do governments. Reflecting a growing awareness of this impact, leaders of international business increasingly accept responsibility for the environment. Moreover, business helps develop cities and human habitats which are ever more urban; therefore, exploring the urban dimension of how business manages the environment is important.

Arguably these changes are redefining the nature of business development and success. Investors are beginning to ask: If we invest now, can we anticipate that the business will be around long enough to recoup our investment? Is this investment sustainable? When business's environmental liabilities become as important in this calculation as its financial ones, a new conception of the international corporation emerges and is directly linked to heightened corporate sensitivity to environmental impact.

With this agenda, opportunities for businesses abound. Since 1972 and the first Earth Day, successes coming out of the last twenty-five years of business change in the environment have seen a great array of new companies and professions. Among all industrial sectors, the environmental services and equipment sector has grown more rapidly than any other during the period 1985–95, approaching in some years an expansion rate of 30 per cent.

Paradoxically, doomsday forecasts, such as those presented two decades ago by Donella and Dennis Meadows and their colleagues in *The Limits to Growth* (1972) help business change.[2] They place the most important critical and radical messages on the table and businesses can respond by developing a positive view toward change.

How to help business actually trigger the complex internal changes in culture needed to accommodate major alterations in the surrounding political-social setting of business is one of the key targets of this book and the research programme behind it. Business opportunities involved in taking long-term commitments to environmental responsibility far overshadow potentials for failure. Whether reform of business practices will save the global environment remains unknown. Without doubt, the combination of increasing population, industrial activity and travel is adding to the biospherical burden of greenhouse gases. Climatic change, it is now agreed, is at least in part triggered by this pollution of the earth's human habitat niche.

What is certain is that without business reform, no stabilisation of the biosphere will ever be possible. Therefore, without supposing that business reform promises any general solution, it will at least provide some degree of progression that coming generations can build upon.

PLAN OF THE BOOK

Part I of the book emphasises the macro-character of business in the planetary environment. Summarising recent developments since the 1980s, Chapter 1 surveys the state of corporate environmental management. Chapters 2 and 3 show how business innovation and a growing awareness of corporate responsibility in progressive companies, including those in developing countries, have affected environmental practices. Chapters 4 and 5 describe the new political ecology of global environmentalism, expressed through environmental diplomacy and the emergence of powerful, entrepreneurial non-governmental organisations.

Part II surveys the local and national focus of business thinking about corporate environmental strategy. This area shows how specific corporate and government tools are being developed to convert the ordinary procedures of business and government into tools for environmental sustainability. Chapters 6 to 9 focus on corporate innovation and response to environmental sustainability in the metropolitan, urban and national context. This section includes a discussion of personal development of managers and entrepreneurs which can be shaped in such a way that they achieve business contribution, security, and personal growth in an ethical framework. Business and the environment are no longer seen as antitheses but rather as interdependent and sustainably symbiotic. Part III presents a vision of ethics and of the future. A detailed examination of the social and environmental dimensions of business ethics comprises Chapters 10 and 11. A presentation and discussion of two contending paradigms of globalisation concludes the book.

OVERVIEW

The 'business and environment' strategic conversation now involves a large global audience. Environmental strategies are infusing the business dialogue. Moreover, in many successful companies, corporate change is linked with a view of how to succeed effectively in altering the values of corporate culture.

The business/environmental strategy field is becoming integrated with other fields of business studies, including strategic management, marketing, finance and human resources development. The market for such thinking is rapidly expanding, as 'traditional' industrial nations seek to join the post-industrial global revolution in business. Even large developing countries such as China, with its rapidly expanding economy and with relatively little experience in coping with environmentalisation procedures for business, have begun to embrace the environmental agenda.

International research, drawing on data from both sides of the Atlantic as well as from the Pacific Rim, shows that global and international businesses are capable of becoming more effective in producing both business and societal benefits. Critics argue that corporations are not going to change their stripes or rebuild the basic instincts which have made them the most successful global institutions we have. From their point of view, the tiger of capitalism has been unchained, and world capitalism, whether controlled by public or private share ownership, provides the major force for world development. Moreover, industrialisation has meant, and unhappily means today, despoliation of the natural and man-made environment. But we believe that this perspective is based upon an increasingly obsolete view of capitalism which fails to take into account the rapidly changing nature of corporate responses to environmental degradation.

Recent changes in the economic structure of India, China and Russia, if nothing else, indicate how far-reaching the corporate model has become. It is not 'free enterprise' that is winning, but corporate enterprise, with public stockholdings, heavy government involvement and regulation, global equity flows and liquid financial markets. If successful, a vigorous public and governmental regulatory apparatus, both national and transnational, will emerge to provide the framework for world development and environmental regeneration for coming generations. Corporate capitalism, this world-wide mixture of enthusiasm, belief and wealth, will be transfigured over the next two generations to produce societal as well as economic and financial benefits.

Just as the quality of government and public administration has improved since 1950, so has corporate governance and management. Business managers are more professional, more corporate and more developmentally focused that at any time in the history of capitalism. Increasingly they come to their business careers from a university education which has exposed them to history, art, humanities and science. Business managers are more aware of their responsibilities and liabilities than ever before. And governments are more willing and able to penalise corporations that get their responsibilities wrong. What is the nature of this change in corporate thinking?

Protecting the global environment: the new orthodoxy

By the end of the 1990s, the fragility of the earth's biosystem and its consequent economic value had become a new orthodoxy.[3] Professor Jane Lubshenco, president of the American Association for the Advancement of Science (AAAS), told the 1997 Seattle congress, 'Protecting the global environment will become the overriding scientific priority of the next century.'

Environmental economics will become a core science linking other sciences to the environment. The economic value of 'free' goods supplied in watersheds, air systems and the seas and ground will have to be priced into development plans. Professor Geoffrey Heal, director of the Columbia University Business School and an economist specialising in environmental issues, symbolised the new unification of business and environmental development. He told the AAAS meeting: 'Preliminary estimates put the aggregate [capital] value of these [environmental] services at or above the GNP of the planet, of the order of $30 trillion.' Cities alone consume huge components of the natural systems and he estimated that the ongoing value of ecoservices drawn from their surroundings is approximately $1 billion per annum. Paul Ehrlich, for decades a leading academic in the global environmental community, suggested at the same conference that natural biological pest control systems have an inherent value of 4 per cent of world GNP, or $120 billion per annum.

But these calculations are in themselves unreal. They do not take into account that in economic terms no substitutability potential exists: if we should irreparably damage the environment we have, no alternative planetary biosphere can be purchased or constructed. So, the generality of economics models cannot be extrapolated other than as a metaphor to the problem of the capital value of the biosphere.

Thus, we are faced with an economics of the absurd which corresponds to Herman Kahn's doomsday scenarios of the nuclear war of some decades ago: pricing what is inherently beyond price is a way of moving our policy systems to achieving the inconceivable. Then it was nuclear disarmament; now it is eco-sustainability.

In all this there is much to be admired and hope is not inappropriate, in part, because corporations are developing a new approach to environmental governance.

A framework for investigating corporate environmental governance

On a world scale, the balance of responsibilities between corporation and community is in continuous change, with government taking on some of the enterprising characteristics of businesses and businesses exercising wider social responsibilities. This change will establish a different way of working between business and its environmental settings: business will soon have to 'own' its environmental performance as a key secondary product.

The desire for any business is to have sustainable successful growth. Traditionally this growth was unfettered. Increasingly the impact of business activity spills over into the physical environment of communities. As a consequence, businesses will be required to demonstrate through their plant and product that their activity enhances, rather than detracts from, the environment.

Recent changes which are likely to affect corporate governance are summarised here.

Environmental standards and national competitiveness

Environmental regulation, it is said by those who reject environmentalism in business, detracts from national and sectoral industrial competitiveness. Proof to the contrary is now emerging. Analysis at the World Bank indicates that countries with high environmental standards were in recent decades able to hold their share of exports and economic activity in competition with countries without high environmental regulations.[4]

Transferring environmental best practice

The transfer of good practice across national boundaries and to developing countries is rapid. Through international organisations, the concept of sustainability will increasingly be extended into a social and ethical agenda. Historically this agenda has been allocated to government, but corporations play an ever-increasing role. Furthermore, electorates are increasingly unwilling to trust government alone with the power and resources essential to meet the enhanced performance expectations for product and environmental excellence.

The social dimensions of environmental management

Community aspirations are not only for sustainability, which is survival by another name. They also want participation and power sharing in key decisions about their future. The corporate system will be prompted through regulatory, voluntary and investment pressures, to meet many of the social agendas of the new century. In many cases they will find, as many outstanding companies already have – for instance, Norsk Hydro in Europe – that community groups are able to build co-operative relationships with manufacturers.

The impact on management careers

Because of these changes, the role of the management executive in the environmental arena is being altered. New university graduates and young business people are faced with a world in which opportunities are proliferating, despite the negative effects of job impermanence. Corporate environmental communications will provide a major focus for development of better understanding over coming years. Managers will innovate improved methods for translating environmental information backwards and forwards across the corporate boundary.[6]

Part I

The Macro-Planetary Process: Business and the Global Environment

Business has restructured itself in recent years to respond to intergovernmental agreements and policy affecting environmental corporate management. Chapter 1 presents two paradigms which are simultaneously emerging to condition the development of world business. On the one hand, a focused responsibility for environmental impacts falls on main international players (the 'environmentalised corporation') and on the other, an increasing reliance on the individual innovator and entrepreneur inside large business emerges (the 'entrepreneuralised corporation'). Chapter 2 shows how corporations are innovating new forms of environmental problems and solutions. Chapter 3 focuses on how corporations, communities, government agencies and action groups are reflecting one another's own priorities and thereby becoming more like one other, and Chapter 4 provides a detailed account of the 'environmentalised' corporation's involvement in international environmental diplomacy and its response to the challenge of treaty-based rules. Chapter 5 presents a case study of an international non-governmental organisation which is structured as a corporate entity and its interaction with both governments and multinational corporations on global environmental issues.

1 The Environmentalisation of International Business

Reality Field 1
Big Business is learning to balance corporate growth with responsibility for environment and community. This process is increasingly driven by creativity and entrepreneurship.

THE FIRST PARADIGM SHIFT: THE ENVIRONMENTALISED CORPORATION

Corporate strategies will increasingly be tilted towards sustainability targets, both short and long term. Many companies and organisations implementing environmental programmes revel in stable marketplaces, with secure products and quasi-monopoly positions. However, international companies must survive in less secure marketplaces. One way or another their capacity to adapt to competition will pivot on their corporate environmental programmes.

Fast-moving markets in consumer goods respond to environmentally-oriented products. Cars will increasingly become 'environmentalised', through both legislation and consumer preference. Thus consumer goods companies will show a strong preference for environmentally-led product development strategies over the next decade, as a way of adapting to intensifying competition.

In the coming decade, as in recent years, intense competition will involve the rapid invasion and take-over of safe domestic markets by better designed and/or better-value imported products, either from within the local region or from producers in burgeoning developing countries.

'Environmentalisation' has become a major strategy for local companies to continue to hold their home markets. A unique selling proposition will involve not only 'local jobs and local prosperity', but also cleaning up plant and product by local producers. If a company takes an ethical stance towards its products, this strategy will form part of the perceived core attraction for increasingly sophisticated consumers. Given the rapid development of environmentalism as an important issue for both corporations and governments over the last two decades, it is now possible to identify the primary features of the environmentalised international business, as well as the management policies which are essential for an environmentally- oriented business.

The environmental revolution and global car exports

The world car market remains the world's largest single manufacturing-centred industry. In this industry we see exhibited all the stresses and opportunities arising from rapid environmentalisation of world industry. As Table 1.1 indicates, Japan and Europe remain major sources of new American car purchases through the end of the century, although the rising trend of the 1980s has been stemmed by a substantial increase in new American cars coming to the market.

If Japan and Europe want to continue to play an important role in American auto sales, they will have to adapt their products to increasingly demanding clean-air targets emerging in various American localities, in particular in California. In turn, these objectives will affect the home markets of European and Japanese car makers. For world car companies, competing internationally depends increasingly on competing environmentally.

The force of the world-wide environmental debate has moved successively from issues of doubt and substantive analysis into issues of compliance, increasing expectation of improved environmental performance, and anticipatory 'pro-activity'. Major businesses are increasingly expected to over-achieve environmental targets, rather than to respond laggardly to legislation. The driving forces are insurance and investment markets, as well as consumer preferences.

A concise statement of the working assumptions about the environment and global change underlying the analysis here can be made, as illustrated in Box 1.1. This 'prologue to corporate environmental strategy' responds to the world-wide consensus prevailing in the

Table 1.1 New car sales in America, 1970–99 (projected)

	USA-made cars (%)	*Japanese-made cars (%)*	*European-made cars (%)*
1970–1979	83	9	8
1980–1989	74	21	5
1990–1999 est.	81	15	4

Source: American Association of Automobile Manufacturing data; percentages provided by Dr Peter Collins.

mid-1990s concerning the understanding, pace and quality of environmental change.

Box 1.1 Prologue to corporate environmental strategy: core syllogism for planetary management and sustainable business: three principles and ten edicts

PRINCIPLE 1: THE PLANET HAS A FINITE BUT UNKNOWN CARRYING CAPACITY

1. All human activity is polluting.
2. Economic activity is the most polluting of all forms of human activity.
3. The planet has a finite capacity to absorb pollution.
4. Therefore, planet-based human economic activity has a finite limit.

PRINCIPLE 2: OUR ABILITY TO PREDICT IS LIMITED BUT IMPROVES OVER TIME

5. As yet there is no generally accepted model of planetary carrying capacity specifying:

- time horizons
- rates of accumulation
- limits of substance accumulation
- recycling and recovery potentials and rates

PRINCIPLE 3: RESPONSIBLE PLANETARY STEWARDSHIP – THE POLICY OF 'NO REGRETS' – LIES AT THE CENTRE OF ENVIRONMENTAL ACTION AND POLICY

6. Responsible governance and stewardship, and the processes to deliver these outcomes, are required.
7. Transnational institutions are appropriate delivery vehicles for such governance.
8. Human wealth creation and planetary sustainability are inextricably interlinked and potentially in conflict.
9. Near-term (5–15 years) action focuses in particular on:

- Reducing pollution output from wealthier nations, in particular
- Developing best available environmentally viable technology
- Transferring best available environmentally viable technology at negative cost to developing nations
- Encouraging and rewarding private and business reinvestment in increasingly less polluting technologies
- Penalising maintenance of high and/or obsolete technologies

10. All sectors, public, non-governmental and private, must invest in long-term programmes to ensure the following:

- more knowledge and better information
- societal and public awareness
- better scientific understanding and prediction
- improved technologies
- tax system incentives to reinvest in cleaner production technologies
- cross-sector co-operation
- international collaboration
- ethical and equitable development of locations and environments
- environmental literacy, awareness and increasing participation in stewardship roles for all sections of global communities.

Corporate environmental strategy: the international economic setting

Companies evolving in the direction of a coherent environmental strategy do so in the context of changing government. The world today is characterised by the interaction of approximately 184 sovereign national states who are members of the United Nations with over 500 international enterprises who are producing and trading on a 'transnational–national' basis. A large number of the *Fortune* Global 500 companies in 1995 had gross turnovers larger than the Gross National Products of the majority of smaller developing countries. Put another way, there may soon be many global companies with turnovers exceeding $10 billion, such that their number will exceed states with Gross National Products of that size or larger. Table 1.2 compares annual turnovers of corporations with developing country GNP, showing the disproportionate weight of corporate finance.

The implications for global environmental governance are numerous. As companies increase their global reach, they will increasingly develop means of overcoming national regulations and controls. While maintaining good relations with regional authorities, they will aim to maximise revenues and minimise tax and environmental liabilities.

A further implication is that global corporations at the top level – the top 100, for instance – will develop means of co-operation to confirm their status and dominance. One method will be through joining together in international environmental voluntary mechanisms and alliances. Because of their reach across continents, most will have more impact on the global environment than national governments.

Table 1.2 The new economic world order: revenues on an annual basis, comparing large multinational corporations and developing countries' gross national product

Rank in Fortune Global 500	*MNCs: annual revenue*	*Annual revenues ($ billion)*	*Newly industrialised countries*	*Per capita GNP ($)*	*GNP ($ billion) (1992 or 1993 est.)*
1	Mitsubishi (Japan)	176	India	310	300
4	Sumitomo (Japan)	163	Indonesia	670	128
5	General Motors (US)	155	Malaysia	3,275	66
7	Ford (US)	128	Pakistan	132	54
8	Exxon (US)	102	Nigeria	89	28
10	Shell (Neth., UK)	95	Peru	950	22
11	Toyota (Japan)	88	Libya	5,410	22
12	Wall-Mart (US)	83	Slovenia	6,330	13
13	Hitachi (Japan)	76	Sudan	400	12.4
14	Nippon Telephone (Japan)	75	Sri Lanka	540	8
15	AT&T (US)	75	Papua New Guinea	950	4
19	General Electric (US)	65	Honduras	580	3.5
20	Daimler-Benz (Ger.)	64	Namibia	1,160	2.3
21	IBM (US)	59	Mozambique	80	1.3
na	Nissan (Japan)	42	Eritrea	77	0.3
na	Nestlé (Switz.)	41			
na	Fiat (Italy)	34			

Note: For comparative purposes, the US GNP per capita was about $25,000 and the UK about $17,000.
Sources: Hunter, *Statesman Year Book 1995*; *Fortune*, July 1995.

The accumulating evidence is clear. Over the coming decades, global companies will inevitably make key environmental breakthroughs. They will find that optimising clean production, distribution and recycling potentials for products will bring commercial success. Higher returns on capital will be evident. Greater social responsibility will have impacts on marketing. Their own market strength will be validated and made credible by good environmental citizenship.

One transitional enterprise undergoing environmentalisation is the British chemical giant ICI. In the early 1990s, ICI, among the 10 largest chemical manufacturers in the world, adopted an environmental policy for its production plants. Implemented over a few years, environmental control and waste/leakage minimisation systems produced substantial improvements in profits through control.[1] Moreover, its plants received fewer complaints about nuisance air discharges than in previous comparable periods. On the negative side, some ICI products which have become less acceptable in the West, including restricted pesticides, are alleged to be marketed in developing countries with less stringent regulations.

However, the trend is clear: such practices will be targeted as unacceptable by national governments. Large companies are increasingly turning themselves into environmentally responsible actors, not only because they are legally required to, but also because they see in the environmental arena a field in which they must keep up with the competition. They also aim to keep ahead of government. ICI sells much of its chemical production to German customers, who in turn have agreed with their governments that all suppliers in the German environmental supply chain must demonstrate that they are environmentally responsible.

Globalisation of investment and impacts on environmental strategy

For global companies, national boundaries exist as impediments to their growth. Such companies are overseen by boards of directors responsible to shareholders and stock market investors concentrated in the global capital centres of London, New York, and Tokyo. These three cities control investments amounting to some 80 per cent of the corporate and business wealth on the planet. While there are many other centres of investment (Frankfurt, Paris, Peking, Hong Kong, Shanghai, Sydney, Toronto, Bombay, Miami, Los Angeles, Chicago, Zurich, Milan and Bermuda) these tend to be specialised in national corporate investment or in particular sectors, such as insurance, commodities or reinsurance.

The problem for environmental strategy, on the one hand, and transnational investment capital on the other is that the environment is a highly localised entity. As well as contributing to national targets for minimising waste and pollution, local businesses affect community environments. There is an argument that global centralisation of investment markets dictates that environmental considerations will be downgraded. This arises partly because the environment is never 'paid for' as are other resources contributing to corporate profits; and also because environmental damage is localised and inadequately reported by either government or business. As Table 1.2 previously demonstrated, the disproportionate weight of corporate finance can be illustrated by comparing the annual turnovers of corporations with developing country GNP.

The quality of information and communications globally is accelerating. The establishment of ethical and environmental investment funds in the last decade has proven successful. New York and London in particular have shown that they can continue to compete for investment markets using criteria which are dependent on improved knowledge of local environmental performance of corporations. This capacity indicates that the traditional centres of investment capital can adapt to the environmental trend sweeping consumer markets.

Key cultural and historical circumstances mean that investment may stay concentrated in Tokyo, London and New York. While share trading and investment may be executed through computers, the actual relationships which underpin stock exchanges are built on trust among dealers. This requires personal familiarity and access among a large number of expert players in the various markets.

Such marketplaces require city regions which can cater for the educational, cultural and recreational needs of tens of thousands of these expert professionals in the financial services sector. The companies and the dealers occupy a unique position in the international marketplace. They have competencies and skills which are not widely distributed and they are at least as valuable as the most talented soccer players and rock stars. Therefore, they can earn high incomes and generate demand for highly specialised consumer goods and services which can be catered for only in large metropolises.

Global companies seek investment from such expert communities of traders. Stock market investors deal on behalf of large institutions and companies – insurance companies, pension funds, mutual funds, investment trusts and other financial management bodies. Some, such as American and British government pension fund management groups, are based in public-sector organisations. However, they function much as do their private-sector equivalents – to invest in a manner which ensures long-term security for their beneficiaries.

Global corporations range from the super-rich German, Anglo-American, and Japanese corporations in oil, engineering and manufacturing to the austerely managed but extensive state enterprises around such activities as telephones, post and transport. In addition to shareholder controls, they each function in response to internally driven corporate systems and may have only a distant relationship with governments.

The desire for any business is to have sustainable successful growth. Traditionally this growth was unfettered. Increasingly the impacts of business activity have spilled over into the physical environment of communities. As a consequence for the first time, businesses must now demonstrate that their activity maintains, rather than detracts from, environmental sustainability. Because they have long-term commitments to look after, investing groups like pension funds and insurance companies are increasingly concerned that companies whose shares they buy are able to maintain a stable position in their marketplaces. Shares bought today should be viable as an investment over a period of 50 to 100 years in the future, as the companies' liabilities to their pension holders and insurance policy beneficiaries collect on their policies.

Increasingly the environmental performance of companies is seen as a key indicator of the soundness of their overall business strategy. If a company is being heavily criticised by pressure groups and is faced with government environmental penalties, its management is not giving high priority to maintaining environmental credibility. In recent years the experiences of Exxon in Alaska and Shell in the North Sea indicate how major financial liabilities follow quickly from environmental neglect. In the Exxon case, damages and costs reached some $9 billion. With Shell, loss of sales reaching $200 million per day resulted during the summer of 1995. Share prices were only marginally recessed, for these are long-established and powerful companies. If environmental evasion had continued, investment groups would have downgraded their recommendations for further investment. Other costs – for insurance and for loan capital, for instance – would have risen, as the companies would be seen to be operating under a more risky management approach.

For the first time, then, large businesses are operating under the keen surveillance of global media. Consumers remote from the environment being damaged have learned to perceive environmental destruction as unacceptable. They are willing to alter their daily purchasing patterns to express their disapproval. Such changes are reflected in the investment centres and bring about financial market responses to environmental depredation. The global village is now under an environmentally sensitised microscope and the nature of multinational business has been altered for good.

Government and the environment

These patterns suggest that consumer and investor power to compel environmental performance is increasing. At the same time, government influence over corporate environmental performance may be decreasing.

Government environmental regulation, many economic think-tanks on the Right have said, detracts from national and sectoral competitiveness. In other words, if you want to be rich, they assert, you have to have industries which pollute the environment. In the short term, there may be some truth to this. However, over a longer term, this notion has been shown *not* to be supported by the evidence.[2] Countries with high environmental standards and strict regulations are able to sustain their world share of exports and economic activity in competition with countries without high environmental regulations (see Table 1.3). They suffer only a slight loss of competitive advantage as environmental costs appear to improve energy and material efficiency. Moreover, environmental costs in the period 1970–90 comprise, even in highly environmentalised economies, less than 2 per cent of total investment.

The analysis uses an index called 'Revealed Comparative Advantage' or RCA, interpreted in Table 1.3. Industrial countries of the First World have suffered only marginal loss in their comparative advantage in environmentally sensitive industries and goods, in spite of having higher costs and greater environmental regulation. There is

Table 1.3 Comparative advantage in environmentally sensitive goods, 1970–90

	1970 index of revealed comparative advantage	*1990*	*Approx. change, plus or minus, over 20 years*
Industrial countries	1.1	1.0	−9%
Developing countries	0.8	0.9	+11%

Source: BSRU/U of Greenwich 1994; original data from Sorsa, 1994, adapted from table 5, p. 9.

no indication of a massive invasion of First World economies by cheaply produced goods from poorer countries with no environmental investment or regulation.

Environmental strategies are generating new kinds of international businesses. Japan's environmental ambitions, as well as the leading developments of the European Community and its top corporations, require a long-term commitment to environmental responsibility. Similarly, through political, legal and investment systems, the concept of sustainability will increasingly be extended into a social and ethical agenda for companies. Typically this agenda has been allocated to government. However, electorates are increasingly unwilling to trust government with the power essential to meet the enhanced performance expectations for product and environmental security.

THE SECOND PARADIGM SHIFT: THE CORPORATION BECOMES ENTREPRENEURIALISED

Because of changes in technology and strategy, the role of the management executive is being altered. While the corporation is becoming environmentalised, new university graduates and young business people are faced with a world in which *both* opportunities and threats in their personal future are proliferating.

The role of environmental sustainability in business is paralleled by other changes. Concepts about work and success now look obsolete. The notion of a single professional training is being replaced with the need for continuous professional development across working life. Business participants are faced with major job switches, often changing countries and sectors in the process.

The notion of a 'terminal qualification' is being replaced by competence-based, rapidly changing stakeholder-inspired qualifications. While general education, in particular business communications skills, is ever more central, so is specialist professional knowledge in a variety of fields. Each technological area is being propelled by its own pool of experts. But all professions and management roles in business are becoming increasingly environmentalist: that is, they are taking account of their responsibilities to the community and the environment and this is becoming as important as financial accountability.

The revolution of environmental responsibility is being generated amid another revolution in career structures. The definition of the corporation is changing. In many of the most rapidly growing sectors, the new international corporation consists of a heterogeneous, often younger, often female, shifting network of autonomous professionals. This network corporation is characterised as a 'learning web'. The key management networks will represent a wide variety of ethnic and racial backgrounds. These fields include entertainment, information technology, software training, marketing, distribution, education, publishing, and health care.

Increasingly manufacturing is being reinterpreted as part of a value chain of customer services within an environmental system. The example that will be dominant in the coming decade is the car, which is being reprogrammed as part of a mobility environment whose technology will become increasingly sustainable.

Key providers in the network may be teleworking from local technology centres. Where networks of sales representatives are involved, there may be no 'central office' facility. Increasingly substantial global businesses are based in widely dispersed telecomputer networks. Their physical environmental impacts are different from and more dispersed from the factory corporations. But for their part, factory-based corporations are dispersing world-wide and downsizing. In advanced industrial regions of Europe, Japan and North America, new plants are being designed for energy and materials efficiency and will recycle waste products to the maximum. Within twenty years, the factory of the twenty-first century will be integrated both into landscape and into recycling systems, with a consequent receding of the problems of site and watercourse pollution. Large 'industrial ecology' factory clusters will be increasingly designed in close juxtaposition, so that their 'metabolisms' form a symbiotic closed system minimising waste, whether materials, energy or pollution. Pollutants are being reprocessed as sources of additional raw materials for other production processes.

As environmentalised corporations become increasingly entrepreneuralised, some additional trends in corporate evolution may be identified.

Careers versus vocations

To enable the environmentalisation of industry, the manager will require a much higher level of professionalisation and vocation than before. Until now a manager needed a career plan, driven by external expectations. Increasingly he or she will need to have a vocation, driven by an internal compass of environmental values, business missions and long-term personal development goals.

Many of the sources for the new corporate designs and models will come from a new generation of small service corporations. Large companies will increasingly be driven by younger managers. Given the need for recent technological training and unlimited energy, managers are often too old at 50 for promotion to senior posts. In a reversal of traditional roles, the over-fifties with suitable pension plans often establish new small enterprises. They become the entrepreneurial 'monkey works', generating innovations and driving forward new ideas for the information economy.

These trends have important implications for everyone at work and in particular for young people. The rising generation of talented professionals and managers needs to embrace rapid change coming from the environmental agenda. Opportunities for early promotion and reward are increasing. Guidelines to optimise one's own chances in this setting need to be developed.

In particular, corporate strategies for developing methods of not only surviving but benefiting from the change to environmentally reprogrammed industries are important. As this pattern is evolving, it is evident that the continuing revolution in business technology is generating new ways of doing business and of personal development in business.

The impact of cyberspace technology on economic dematerialisation

As an introduction, consider the following passage from Doug Rushkoff's *Cyberia* (1993):

> Having turned the concepts of a new math called chaos theory into a working model for reality,... existence [becomes] a field of interdependent equations.... Welcome to Cyberia.
>
> Time seems to be speeding up. New ideas and technologies have accelerated our culture into an almost unrecognisable reality, and those on the frontier tell us that this is only the beginning. The different ways in which our culture is changing can be understood as different aspects of a single renaissance. Inspired by the computer, chaos math, chemicals, and creativity, this renaissance has been interpreted by many as an evolutionary leap for humanity into another dimension. Whether or not this is true, those who can

comprehend the nature of this shift will be better prepared to survive the twenty-first century.[3]

The new information systems revolution taking hold in business requires a new set of sensibilities from managers and entrepreneurs over coming years. This change constitutes a renewal of a culture of change. Innovators and managers who comprehended the direction and quality of change at the time benefited. Younger managers evolved ways of thinking, writing, communicating and acting which succeeded against the post-Second World War approaches in the professions and businesses.

The impact of telecommunications is intensifying this shift. Businesses based on new telecomputing systems will produce phenomenal benefits and impose enormous dislocation on communities unable to adapt. In the meantime, older businesses where the successes of the mid-century period still dominate management thinking will find themselves priced out of the market. Environmentalisation through real-time telecommunications systems is one of the key dimensions around which new business strategies will be based. Cyberspace technology and the business–environmental revolution interact. The radically enhanced power of telecomputerisation and environmentalised global business are interdependent. Indeed the twinned developments will feed on and reinforce one another.

The psychotechnology of innovation and entrepreneurial creativity

While creative lateral business innovation is newly 'reinvented' every 30 years or so, the approach is a recurring feature of industrial development over the last 200 years. Individuals by means of a radical structuring of attitude learn to evoke positive results. Such re-engineering, as Champy and Hammer point out, is a psychotechnology of innovation.[4] A popular 'success' literature continued through the twentieth century, prominently in the USA. For example, Napoleon Hill, as a young journalist, was challenged by the industrialist. Andrew Carnegie, to analyse Carnegie's success system and then to develop a 'science' based upon it. Becoming wealthy himself, Hill produced the approach and worked from 1915 through the 1940s. He wrote some two dozen books, many of which continue to sell in thousands world-wide, and founded institutes and academies teaching the Carnegie success method. His principles focus on the following five steps:

1. Constant attention to attainable goals
2. Daily and disciplined practice of effort which is known to lead to these goals
3. Patience, good humour, and worthwhile human values
4. Love, loyalty, and friendship
5. Loving and supportive self-criticism of ourselves, our families and work teams.[5]

His well-documented definition of success is often quoted: 'The day-by-day achievement of successive, cumulative targets leading to one large overriding goal.' No life which is deemed to be successful can be built in any other way. Whether the place of work is charitable, collectivist, or entrepreneurial, it must be structured by the individual innovator as a succession of attainable goals, which offer cumulative rewards, recognition and personal satisfaction.

The implication for our analysis is this: The centrality of the individual role in building culture and prosperity is part of the next step in world business creativity. As companies take a greater concern for the long-term sustainability of the environment and corporate community impacts, individuals working in collegial teams are seen as essential. In business, the recent revival of leadership theory recognises that in the end *individuals* are the source of economic creativity and wealth.

The first environmental conquest: energy efficiency

Since 1970 success in one area of environmental management, energy conservation, suggests the pace of change that will continue. The development of energy efficiency has been driven by three fears: first, energy is expensive and forms too high a proportion of expenditures, whether for companies or households; and second, it is not in the control of major developed countries' governments in the strong economies of Europe, Japan and the USA, which must import large proportions of their energy requirements. By the 1990s energy prices had fallen back to the historically low levels of the early 1970s, in relative terms. The technical reason is that for every unit of GDP being produced, the energy requirement was about one-quarter of that of 1970.[6]

In the 1990s a third major fear driving the search for energy efficiency has emerged. Widespread awareness of the potentially negative effects of global warming has led both governments and major corporations to improve energy efficiency as part of a broad strategy to reduce ozone depletion. David Nemtzow of the American

Alliance to Save Energy comments: 'The politics have changed. Now, [energy policy] is overwhelmingly driven by the environment.'

IMPLICATIONS OF THE SECOND PARADIGM SHIFT

The analysis pursued here focuses on the real world of what business are doing now and how managers and the public can assist and motivate businesses to do more. The argument is that the earth is still capable of being saved from irreparable damage. Many environmentalists, including systems scientists, like Donella and Dennis Meadows, argue that it is too late to protect the global environment.[7] Other writers, like Greg Easterbrook, argue that the pressures, already building up inside and outside advanced industrial economies, mean that environmental regeneration over the coming decades is, if not inevitable, at least likely.[8]

No one knows the correct answer. What is clearly the case is that changes are afoot. Business, government and the public are willing to accept levels of environmental regulation and control which were rejected twenty years ago. In technical terms, there is the prospect of a paradigm shift, an alteration of the cultural web which ties together the way modern economies work. As Table 1.4 illustrates, we are changing our behaviour, at least in the advanced industrial economies.

Table 1.4 The corporation as negotiated environmentalist and enabler of entrepreneurial activity: the second paradigm shift

	Old paradigm	*New paradigm*
1. Biophysical underpinning of business activity	External to corporation.	Internal to corporation.
2. Stakeholder analysis	Management plus shareholders.	PLUS: customers, employees, community interests.
3. Maximum forward planning horizon	10 years.	100 years.
4. Audit rubric	Financial.	PLUS: human resources, environmental, technological.
5. Environment valuation	A free good to be consumed at will.	A high-value good to be conserved and renewed.
6. Role of the individual	Working in conformity to corporate culture and according to established norms, e.g. 'job description'.	A self-enabled innovator and entrepreneur, taking action on behalf of the company and continually redefining her role and responsibilities in positive negotiation with the organisation.

Rapidly developing economies, like China and India, are acquiring with advanced electronic technology and accelerating consumer demands, higher environmental expectations. These ancient societies undergoing radical modernisation will not necessarily tolerate a lower standard of environment as a price of prosperity.

So, there is a likelihood that radical change is under way. Whether it is fast enough or big enough is the only argument. Business on the one hand attempts to control its markets. But on the other, it also must respond to changing demands from all its stakeholders. This book is focused on how this response is being shaped and, as well, pursues the line that the response can be improved. Thus, many suggestions about better procedures are positioned in every chapter to illustrate that business and the coming generation of international innovators can sharpen these efforts.

With respect to the broader debate regarding the condition of the planetary biosphere (fragile or robust) and the subsequent division of environmentalists into optimists and pessimists, we must hope that the optimists are right, even if we have become convinced that all efforts will prove inadequate. The eco-realist argues that, despite the debates, there is also broad area of agreement. We do not know the future, nor are we certain about the direction or rate of change in the biosphere resulting from human activity. With the continuing expansion of human economic activity, the capacity of the biosphere to absorb additional pollution will inevitably be tested.

Indeed, nature itself alters the biosphere in calamitous fashion. It is clear that on average about every two centuries the biosphere has to absorb the results of massive volcanic explosions – Krakatoa in the 1890s, Santorini about 1000 BC. Mt. Penatubo in the Philippines erupted in 1991 and its stratospheric fallout limited sunlight reaching the planet for some three years after. This phenomenon causes a substantial decrease in sunlight for one to two years in many parts of the planet. Whole economies are blighted. History shows that farming has suffered repeatedly and famines have resulted from natural events in the biosphere.

It is conceivable that human activity can emulate the effects of volcanoes. If volcanic eruption and cumulative air pollution coincide simultaneously across the globe, it is not illogical to believe that severe losses of sunlight could persist for several years. The consequences for 8–10 billion human beings of the twenty-first century in terms of food and energy requirements can be imagined.

We must operate in this generation with a view that the art of managing planetary ecology will have to be developed by coming generations. We are founding a new ecoscience and a new craft of planetary ecology. And the battle is fought at the micro-level, in every business and household.

This art, now in its infancy, is reflected in responses to our periodic environmental scares and panics. Such issues as DDT in the environment, acid rain, deforestation, desertification, depletion of the ozone layer – all have inspired more governmental regulation, both national and international. Sometimes the responses have not been justified by later evidence.

However, this imprecision is inevitable in a new global management process. What is important is that progress is being made in this art. Over the next 50 years, we must be able to know that our best efforts to anticipate and control disasters of such magnitude and depth have been made, or we truly will have become ignorant agents of Kali, the Hindu god of destruction. Global corporations' approach to the new management challenge is the subject of the next chapter.

2 Global Business Innovation and the Environment

Reality Field 2
The emergent global system requires management approaches combining local sensitivity with planetary vision.

INTRODUCTION: THE CHANGING CORPORATE CULTURE

In building cities, business links together communities, families and the planetary environment. By designing environment structures which are responsive and re-programmable, corporations offer the most successful tested method of forming habitats and urban regions offering sustainable lifestyles to the expanding human population.

Contribution to communities, global and local, will be increasingly essential for sustainable business presence over coming decades. The interests of shareholders, the legal owners of *public companies*, are seen as linked to the interests of other stakeholders. Moreover, the long-term interests of shareholders may well mean reducing their short-term profits and dividends. Shareholder governance means that the corporation is run solely for the financial benefit of its shareholders. By contrast, *stakeholder* governance requires corporate management to balance the interests of all parties concerned: shareholders, employees, managers, customers, suppliers, neighbours, communities and government agencies.

The evolution of shareholder governance into stakeholder corporate governance is proceeding at speed. In the continental countries of the European Union, especially France, Germany and the Netherlands, stakeholder capitalism has been extensively studied and developed since the Second World War. German corporate law requires corporate governance systems with Workers' Councils directly advising the main board. In Japan there is a long tradition of responsibility to employees and to communities.

In the environmental field a notion has developed that antipathy exists between on the one hand potentially polluting companies and on the other local communities and interest groups. To offset this orientation, increasingly corporate governance includes many of the groups who have come to see business as an 'enemy'. As faith in government has declined, the belief in markets has expanded. Corporations – public, private, not-for-profit, local, state-owned and international – are the major players in world marketplaces. Companies driven by the entrepreneurial spirit are still seen, of course, but they most often evolve into corporations of tomorrow. For these reasons, they are taking on some of the trust and responsibilities which big government in mid-century carried. As noted in other chapters, government is becoming more entrepreneurial and decentralised like business; this has implications for the environment as well.

Companies, particularly large multinational enterprises, are worlds unto themselves. Because they involve tens of thousands of employees in many countries, they become 'political states'. Within each company, specific beliefs become deeply built into their practices. As in tribal cultures, rituals and myths permeate and differentiate the company and indeed are often seen as part of the reasons for its competitive advantage.

In coming years, deeply ingrained rituals and myths in most international businesses will be challenged by many changes. The rising predominance of environmentalised corporate strategy is one such area, along with changes in telecommunications technology.

The international oil companies, some of the most powerful and secretive of corporations on the planet, are beginning to show the way ahead. The Seven Sisters of big oil – British Petroleum, Exxon, Mobil, Shell, Elf Aquitaine, Aramco, Esso – each has a different approach to its missions as well as specific rituals and myths. On the whole, large oil companies have rejected for many decades the problems of their environmental impact. Their traditions are embedded in the 'Macho Management' of nineteenth-century robber barons, even while they struggle against one another for leadership in world petroleum markets.

In the 1990s, first with Exxon and then with Shell, fundamental flaws in the underlying paradigm of rejective-environmentalism and its accompanying myths, are being revealed. Consumers and 'the public', once an amorphous mass market capable of being manipulated into appropriate buying patterns, has sprouted powerful non-governmental organisations (NGOs), like Greenpeace and Friends of the Earth (see Chapter 5 for a detailed examination of NGO corporate cultures). Many environmental activists, as well as newly formed international agencies such as the World Trading Organisation (WTO), were beginning to realise that if multinational corporations are operating globally, then environmental inspection, investigation and regulation must also become global. In response to these changing trends, corporations have begun to innovate in response to growing concern for environmental sustainability. We have selected four specific case studies to demonstrate the innovative nature of corporate responses to the challenge posed by environmental deterioration.

CASE: COMPANY ENVIRONMENTAL MANAGEMENT IN THE US PACIFIC NORTHWEST[1]

By the turn of the century, most leading North American and European companies will have established in each of their marketplaces an 'environmental position'.[1] Companies are increasingly committing themselves not only to reactive compliance, as illustrated in Table 2.1, but also to integrative alignment of business strategy to principles of environmental stewardship.

Table 2.1 Pacific Northwest companies (N = 36)

Group	*Term*	*No. of companies in sample*
1	Compliance	6
2	Alignment	10
3	Expansion	14
4	Integration	6

Source: Adapted from Scallon and Sten, 1996.

Senior managers in one sample of northwest American companies saw environmental issues in a wide variety of ways. Some see environmental expectations as obstacles. Others see the environmental issue as one on which their company can build opportunities. Where obstacles are emphasised, companies spend time on ensuring compliance and avoiding potential problems.

By contrast, managers who emphasise opportunities focus on the search for such opportunities. Where such opportunities are yet to be discovered, these managers aim to 'do the right thing' anyway, as a reflection of core values in their corporate cultures.

Table 2.2 A typology of corporate environmental strategies

Corporate type: Comparative dimension	*Complier*	*Aligner*	*Expander*	*Integrator*
Regulatory perception	Keep up and comply.	Keep ahead and comply early.	Work to influence regulations in a positive way.	Work to influence regulations in a positive way.
Risk-reduction activities	Focus on least-cost alternatives in reducing environmental risk, e.g. BATNEEC* analysis.	Recognise that voluntary compliance programmes may offset later and greater expenditure to catch up.	Systems, e.g., on going environmental audit to detect potential improvements in environmental performance technologies.	Environmental issues and risks are structured at heart of corporate plans.
Opportunity search for means of improving environmental performance	Given little prominence.	Unwilling to invest in environmental measures without demonstrable early returns.	Develop environmental activities involving, e.g., suppliers, customers, community liaison groups.	Systematic environmental isation of all major corporate functions.
Prominence of environmental issues with decisions, stakeholders	Minimised.	Avoid non-mandatory environmental programmes.	Have in place a broad environmental system and experiment continuously to improve it.	Value environmental programmes without requiring them to break even.

* BATNEEC (Best Available Technology Not Entailing Excessive Cost).
Source: Adapted from Scallon and Sten, 1996.

Table 2.1 indicates that among the 36 prominent companies Scallon and Sten surveyed in the Pacific Northwest in the early 1990s, four types of corporate environmental governance cultures can be identified. Each has specific behavioural aspects as indicated in Table 2.2.

A further important aspect of this research is that location in the typology does *Not* correlate with any of the following dimensions:

- size of company
- location in supply chain
- industry type

In other words, large and small companies were found in all the categories. Retail, manufacturing and distribution companies were found in all categories. And all types of industry were found in all categories. Represented in this group were the following industries: aerospace, chemical, clothing, electronics, food processing, grocery distribution, heavy construction, health care, manufacturing, oil refining, port authority, retailing, financial services. What is clear from the survey is that a specific, environmentally sensitive culture is emerging in three of the four corporate types. As companies move toward environmental sensitivity, their strategies may be identified as aligner, expander and integrator.

CASE: INDIA AND THE INTERNET-DRIVEN URBAN EXPLOSION IN NEWLY INDUSTRIALISING COUNTRIES

Newly industrialising countries (NICs) are experiencing the impact of world electronic telephonic and wireless networks. Among these countries, India provides the most startling case. As recently as the 1970s, India was seen as a permanent international dependent, with deeply entrenched poverty, an increasing population of destitutes, and a cultural miasma biased against economic and technological development.

By the late 1990s much of the poverty continued. However, in a number of urban regions, investment and systematic cultivation of key industries had provided the basis of a boom with growing momentum. India was becoming self-sufficient in food and steel, two of the traditional industrial sectors. Remarkably however, India had educated a whole generation of electronic engineers and scientists. It also had a range of state incentives for banking institutions, a creative regional development agency and a large number of wealthy family-based businesses with international networks and a cosmopolitan approach.

Two areas of change provided the key factors for this success. One was the large number of trained people with low salaries, compared to international standards. Driven by the examples of American university expansion, similar increases in higher education were put in place in India. This produced a large number of technology graduates with only a few outlets for their employment, and salary expectations were relatively low.

At the same time the international telecommunications networks – both wired and satellite-based – were being put in place. Inexpensive –

and non-polluting – approaches to air-conditioning and new industrial estates were making the tropical Indian climate viable for concentrated indoor working, just as it had in Florida and the American South three decades earlier.

With the development of local technology universities, it had become a successful centre of advanced software writing and database downloading.[2] As in the airline industry, specific database industries were increasingly managed in real-time conditions. Global simultaneity was becoming possible in such hitherto mundane areas as medical billing and record keeping. An important step was the Indian government's decision to ease admission for foreign software companies. Thus, by the late 1980s, cities such as Bombay, the leading industrial centre, were already experiencing acute land shortages due to the influx of those seeking employment. Government policy designated Bangalore as the centre for software and technology development and it became known as India's 'silicon city'. It has made India in the 1990s the second most successful exporter of software, after the USA. In 1995 overseas sales increased by 60 per cent, to around $700 million. For multinationals like Texas Instruments, Motorola and Hewlett Packard, this has proven a successful point for entry into the huge market of the subcontinent.

Environmental impact of global technology booms

The environmental consequences interlink with the industrial wave of dramatic growth in *Bangalore*. Thousands of electronic screens blink in dozens of companies focused in business parks in new buildings around the edge of city. Hectares of new concrete cover what had been agricultural fields. Business parks and technology centres are proliferating to respond to the demand for high-quality air-conditioned offices to accommodate the huge rate of business development.

Until industrialisation, Bangalore was a sleepy southern town, a retirement area for civil servants and the well-to-do. This tranquillity is gone, as today the average employee is 29, single and highly paid by Indian standards. Such chains as Kentucky Fried Chicken and the British department store, Littlewoods, have opened outlets.

Between 1990 and 1995, the city's population doubled from 3.8 to 7.5 million. The narrow roads of the old city carry some 850,000 vehicle movements per day and are permanently congested. The city has been unable to increase transport resources, relying on 300 traffic policemen and 6,000 road sweepers. These are assisted by a network of senior citizens who volunteer as traffic wardens. The airport, designed for 700 people, by 1996 bulged with over 6,000 passengers every working day.

Consequences for land values have been dramatic. Many small owners have sold their holdings and left the city. Poorer families cannot find accommodation and are being forced to leave the centre. Property speculation is rife.

Energy supplies are grossly inadequate. The technology companies build in their own power-generation supplies and have their own sources of water and energy. The city-wide system experiences shut-downs for 3 to 6 hours a day. Voltages are too low to run most appliances. The urban environment reflects within a few kilometres every condition, from pre-industrial subsistence farming to high-tech global telecomputing.

Local planners admit they were caught off guard by the rapidity of development. They now think in terms of developing satellite towns to cater for the specific needs and opportunities of technology firms, at distances as far as 60 miles from the centre.

A global start-up enterprise: HealthScribe

Apart from the American multinationals, a number of new local enterprises have been formed. One such company, HealthScribe, was originated in 1994 by Akasha Ellis and Karan Kaur Khalsa, who are both children of American converts to Sikhism. Goldenberg notes:

> The couple train operators to transcribe medical notes dictated by doctors in the US and sent to India as voice mail at the end of the working day. Surrounded by books on anatomy and medical dictionaries, the operators then send the rambled mutterings back as complete medical records, enabling the doctors to submit a speedy claim to insurance companies.[3]

The business has doubled each year of the last two years and plans include the appointment of 100 additional trainees. In countering accusations of exploitation of Indian conditions, Goldenberg quotes Mr Ellis, co-founder of HealthScribe: 'We are giving an education, we pay for the education, we transport them, we feed them and we give them a better salary than they could get anywhere. And they get a chance to learn about technology.'[4]

Conclusions: the Indian conundrum

For India, it is clear that the development of electronic industries is producing some of the benefits of 'leapfrogging'. That is, the Indian economy is being driven forward by specific developments in a few locations, establishing innovation centres which then radiate back into their hinterlands. At the same time, the consumer problems of Western economies are also being established: excessive packaging waste: environmental spillover with the threat of pollution.

In a sense, the Indian economy is beginning to look like an advanced Western economy, except that the proportions of workers in the various sectors are very different. In the USA about a quarter of all households are outside the advanced economy of affluent industrialism. In India this proportion is about 80 per cent. But the proportion is shifting. Increasingly over coming decades, this huge economy once regarded as irredeemably 'backward' will join the rest of the world's mixed economies of affluence and neglect.

CASE: ENVIRONMENTAL CRISIS IN INDONESIA: THE GRASBERG MINE IN IRIAN JAYA

Global mining concerns have been developing tropical sites in newly industrialising countries for a century and a half. This process was carried out in much the same fashion decade after decade, ignoring the many changes being imposed on industrial concerns in the advanced industrial economies.

The old tradition was well established. Combining with the governmental officials, the mining companies, in alliance with local governments and local mining interests, were funded by London and New York investment groups. They identify large deposits of minerals in backward regions. Sometimes, operating in secrecy, they failed to provide local leaders with the details of the development process.

In this 'traditional' approach, development occurs in massive and sudden excavations. Mine tailings are not only dumped into the environment, polluting both rivers and coastlines, but they also form permanent alterations to the landscape. Chemical outputs leach into the ground water and devastate surrounding areas. A few local people may be employed, but in general the workforce is imported and kept in dormitories with little access to the outside world. Within a few decades, the deposits are worked out and the sites, often covering many square miles, are abandoned, leaving giant, permanent, and open scars on the landscape.

Local ecologies are debased and in some cases ruined. Local communities are left with spillover of pollution from the air and streams. A range of social and medical problems often traumatise local communities for generations afterwards. In some cases, the development process attracts secondary settlers, who intrude into traditional farming and hunting, forests and fields, debasing and destroying traditional economies and culture.

Mineral wealth is stripped from the natural environment, with almost none of the financial benefits being used to improve human or environmental resources of adjacent communities. Profits are remitted to global investors and to the national governing elite based in capital cities. Local communities are corrupted or destroyed and any possibility of using the mineral resource for local development has been obliterated.

This pattern has persisted since the development in the mid-nineteenth century of international mining consortia, initially in Europe itself, then in Africa and in the Americas. In the twentieth century, the pattern of mining exploitation has been particularly visible in Latin America, Africa, the former Soviet Union and in the Pacific island nations. More recently, the pattern continues in the republics which were part of the former Soviet Union.

Since the mid-1980s, a dramatic case of the worst practices fusing together has been presented in Irian Jaya province of Indonesia, as reported by *Financial Times* correspondent, Manuela Saragosa.[5] Freeport, Indonesia is part of the US mining multinational, Freeport McMoran, Copper & Gold. The mine at Grasberg had become one of the largest copper, gold and silver mining operations on the globe.

Like many other overseas corporations, its operations are insured by Overseas Private Investment Corporation, a US government agency. In 1995, OPIC withdrew $100 million of political risk insurance, citing the mine's 'substantial adverse environmental impact'. This dramatic decision was a direct reflection of the work of pressure groups, such as Greenpeace, drawing attention to the Grasberg mining conditions. Reacting to the withdrawal of insurance, the environmental auditors were commissioned by the Freeport company. Working in the remote province of Irian Jaya, the team was drawn from Dames and Moore, a leading American environmental consultancy. Responding to the pressure of US government agencies, Indonesian authorities were keen to demonstrate their concern for and willingness to take action on the environment. Thus, they also became co-sponsors of the environmental audit.

The environmental auditors noted that Freeport Indonesia was 'slow to meet its environmental challenges, adopting a reactive rather than a proactive response' to environmental challenges. Disposal of tailings from the mine into rivers was identified as the most 'visible, unsightly and critical issue'. Although non-toxic, the tailings were 'unwelcome' to local tribes living in villages around the town of Tembagapura, the main local market town.

Responding to concerns of international organisations, the company made efforts toward improving environmental management. The auditors note that current plans to establish a tailings storage area using levees represented an 'acceptable solution'. In addition to the initial cost of $25 million, a further $12 million per annum will be required for maintenance of the levees. The auditors further recommended that Freeport establish a permanent environmental advisory team, with independent members. The aim here is to re-establish 'an atmosphere of trust'. The auditor's report notes that local environmental damage also emanates from other sources, in particular:

- extensive logging operations in the local rain forest, a direct result of government encouragement;
- transmigration policies of the Indonesian government, giving incentives to incoming populations coming from overpopulated Java.

The auditors recommended as well:

- a study of potential future liabilities from past neglect of waste management systems
- greater commitment to regional biodiversity of flora and fauna, through research on fauna and flora biotic bases
- formulation of a detailed mine closure plan.

By the late 1990s, international attention, created through the efforts of a globalised non-governmental organisation, such as Greenpeace, has given credence to the notion that traditional practices of mining pillage and abandonment cannot be sustained by the international insurance community. Both investors and insurers are unwilling to tolerate the financial risk and moral opprobrium which has become inevitably associated with the old ways. At the end of the twentieth century, the reach and instantaneous nature of global communications has fundamentally altered the means by which companies retain exploitation rights.

What can be said of the future? It is conceivable that the Freeport decision of the late 1990s signals the beginning of the end of the centuries-old tradition of 'rape and pillage' multinational mining. Never again will the old practices be tolerated with such cavalier disregard for their environmental effects. However, within national boundaries and driven by authoritarian governments, it is conceivable that such practices will persist for some time. China, for example, has current mining practices which are internally driven and which fall under the old category. Some commentators say that if prison labour is engaged in the mining operation, these practices may well constitute a violation of human rights.

Despite the pessimism induced by retrograde practices in a few countries, the example of the Grasberg mine demonstrates that international non-governmental organisations, the media, corporations and national governments, working toward similar objectives, can take effective remedial steps toward protecting the environment. Those steps contribute importantly to the move away from old management practices and toward the adoption of new ones. The impact of three influences may well be inducing environmental 'villains' and corporate 'bad citizens' to realign their management practices in accordance with the new environmental ethic:

- willingness of government agencies dealing with international trade issues to subject overseas investment projects to an environmental impact assessment before operations can begin
- the increasing capacity of government agencies in developing countries to take an assertive view of environmental performance by inward investors as well as by national firms
- instantaneous global communications of environmental failure.

CASE: A TWENTY-FIRST-CENTURY BUSINESS PROPOSAL: A GLOBAL ENVIRONMENTAL TELEVISION SATELLITE NETWORK?

As the previous case illustrates, the global environmental network, driven by satellite television, allows us to see how the worst practices of corporations can be brought into line with the new environmental ethic. Similarly the power of broadcast media in alerting millions of people to changes in the environment has been demonstrated through the move away from CFCs and the Shell North Sea episode. As the authors of one global conceptual plan observed:

> If we can afford to have national television networks focused on high quality arts and social-economic commentary, like that of Public Broadcasting System USA, BBC 2 television, and France Channel Dixie, there should be a satellite/cable world channel capable of delivering to a world-wide audience, through subscription and fair-trade commercial activity, incisive and critical analysis of environmental events in real time, particularly as there are concerns regarding business impacts on earth sustainability.[6]

A proposal to establish such a network has been under discussion since the late 1990s. The Global Arts Network (GAN) reflects the enormous success of continent-wide and world-wide media campaigns during the 1980s. A British-centred campaign in the mid-1980s raised some $300 million for famine relief in Africa and was repeated in two successive years. Television screening of the pop-star concerts world-wide was a major source of revenue, assisted by the contribution of major entertainment artists of the era.

Strategic business planning of the proposed Global Arts Network takes the arts experience into a permanent global network focused on issues of human welfare and environmental sustainability. A range of forces and key resources are converging to bring into being a new global television network. The establishment of a world-wide business at this level is projected to require $200 million in available capital.

The mission of a global corporation

The mission of GAN is conceived to operate as a global satellite/cable television network, accessible to 75 per cent of humanity and driven by ethical and spiritual values emphasising global sustainability, social equality, trans-cultural entertainment, and human cultural development. Programmes will be presented in English/American, sub-titled as appropriate, in a 24-hour sequence of high professional quality, to include titles drawn from, among other subjects:

- popular and classical music
- news analysis

- performing and visual arts
- business and economics
- history, anthropology, and social commentary.

Programmes will draw on world sources, recent and historical. New programme series would tap the huge number of under-represented film makers and writers.

The background

GAN, when launched, must expand in a matter of months from a loosely-knit extended world seminar into an appropriately structured global television network involving hundreds of contributors and thousands of communities. Planning and preparation require, thus, a period of rapid intensification and formalisation, so that infrastructure and programming are firmly in place at the point of going live. The Foundation Fund of $10 million makes possible the process of design and delivery.

Justification for the project

With the collapse of centralist economies and the marketisation of nominally socialist countries, most governments have accepted that the functioning of international and local markets provides the best way forward for human development and progress. However, there are continuing grave concerns about social inequity – for women, for the poorest countries, for the least well-off in the rich countries, for ethnic minorities in all countries. Environmental resources – air, water, the seas – are not 'priced' in the new market economy. At this time there is no global media voice which explicitly addresses these issues and presents research concerning improved working of markets, political development, human rights and economic improvement.

While the rule of international law and best ethical business practice is still unclear for many areas of the globe, an increasing desire to expand human rights and fair business approaches into all international trading nations exists. No broadly available media source reaches the whole of the human population with news concerning developments in legal affairs, corporate law, and human, occupational and consumer rights.

The information economy, emerging in recent decades in the rich countries, is making available a rapid economic development process for every part of the global economy in coming decades. Relatively few communities on a global basis understand the opportunities and problems for them in such a process. A powerful tool for such understanding is satellite/cable TV. Although the business channels intermittently present macro-economic projections and discussions, as yet no channel addresses systematically the issues of local economic and community development.

To date, the cable/satellite industry has been focused on traditional media products with high commercial value – news, sports, business, pop music, films and repackaged TV series with consumer visibility. In a few cases, novel products have emerged such as the music video. A huge potential demand for additional innovative media products with artistic, musical, professional, environmental, social and cultural-historical content has so far not been addressed by the established industry. In addition, the many musical hits created in the pop-music industry in recent decades have created a generation of successful artist-entrepreneurs and allied production companies with unfulfilled aspirations to contribute to world cultural, political and environmental development.

Media artists from the pop world want an established medium in which their commitment and concerns can be deployed for wider human development purposes. For a majority of the high-income households world-wide, existing global channels are not seen so far as offering anything of value compared with terrestrial media available locally. As the GAN plan points out, the total global population with household discretionary income commensurate with access to satellite/cable TV, and with values supportive of social and environmental responsibility, is 800 million persons, with a total annual household discretionary spending amount of sixteen trillion dollars).

The advent of inexpensive satellite-based communications will transform the logistical management of business and government in many countries in the next decade. Interactive satellite-based telecommunications has significant implications for the media, entertainment, distance learning and public information sectors as well. A global media network should address *the technological development of satellite two-way datacoms*, which will allow the global involvement of hundreds of millions of households in learning systems, polling, commercial purchasing, and other forms of interactive involvement with suppliers, government agencies, non-governmental organisations (Greenpeace, Amnesty International, for example) and citizens of other nations.

Business is becoming more systematic in creating important new products. Increasingly innovation is coming from smaller businesses which, for the first time, can operate on a global scale with small amounts of capital. The internet is the first global media resource able to reach this global community involving tens of millions of individuals, but no global broadcast media network reaching this large and affluent audience has yet been put in place.

The global environment is thought to be under long-term threat from toxic and pollution waste being released into the atmosphere and oceans through human urban and industrial development. Wealthier nations, it is feared, will export toxic waste and outmoded polluting technology to newly industrialising countries. A broadcast medium investigating and globally reporting ecological risks and improved environmental management technologies will find a world audience. As corporations expand into global strategies, so too will public demand for global real-time information about environmental performance of large companies and government agencies.

Summary

The GAN external audit shows that:

- no global medium yet addresses working markets, political development, human rights and economic improvement potential
- no global medium source is reporting on legal affairs, corporate law with an orientation toward human, occupational and consumer rights
- no global medium systematically addresses local economies and community development
- no global medium has tapped into the resources offered by the unfulfilled aspirations of pop musicians to contribute to world cultural, political and environmental development
- no global medium directly and systematically discusses business innovation which can have positive benefits to the environment and to community development
- no global medium investigates and reports regularly on ecological risks and improved environmental management technologies.

GAN identifies a gap in the global media market for a channel offering a wide range of 'info-tainment' products targeted at the higher-income households in all countries. Further commercial opportunities for interactive product marketing of such sectors as green label, fast-moving consumer goods and a range of leisure and entertainment products exist. Whether or not the GAN proposal succeeds in its objective to establish a new global channel, it indicates the extent of innovation which is occurring in the nexus between business and the environment in the late twentieth century.

3 Global Trends and the Corporate Response

Reality Field 3
Identifiable global trends are emerging which will unify the dialogue between business and the environment.

TEN MAJOR TRENDS FORMING THE TWENTY-FIRST-CENTURY INTERFACE BETWEEN BUSINESS AND THE BIOPHYSICAL ENVIRONMENT

1. Scarcity

Scarcities – of space, resources, access, and food – permeate the new century.

Implication: Market-based systems will be used to ration these scarce resources. Large global networks of businesses will be the typical means of running such markets.

2. Competition

Competition among trading nations and national economies will become even more intense.

Implication: National communities will give priority to their international trading capacities.

3. Smaller or less effective government

Governments are becoming either smaller and less powerful or more complex and less effective.

Implication: Recent demands for governments that are smaller, more efficient and less costly will give impetus to governmental initiatives which focus on partnership arrangements with business. These partnerships will strengthen the ability of government to influence business and vice versa.

4. Global business networks

Globally organised business corporations are becoming the dominant international force.

Implication: Increasingly corporations will be operated as semi-autonomous business units, networked into strategic alliances. Being dominant, they are also becoming more responsible on issues of equity and environmental ethics.

5. Sustainability

International business is acquiring a strong commitment to achieving sustainable development.

Implication: Global corporate networks will be increasingly open. They will be more transparent and will communicate with less secrecy with their external stakeholders.

6. The global third sector: voluntary organisations

A recent global third sector, made up of powerful international voluntary organisations, is emerging to counter and complement the power of business and government.

Implication: The third sector is a network of philanthropic enterprises, whose major commodity is information, knowledge and the emergence of ethical systems in trade and development. Because much of it comes through television, it can be categorised as 'infortainment' – material that is valued both for entertainment and information purposes. Greenpeace and Oxfam by the late 1990s are in themselves autonomous and self-managed forces for global systems development. Business needs the third sector, to make certain that competition among giant firms is carried on in terms that safeguard the planet's sustainability. Global corporate networks will invest in, sustain, and encourage third sector enterprises, in order to improve its anticipation of potential ethical and environmental dilemmas and self-inflicted damage of future markets.

7. Re-targeted government

Government, the first sector, will find new methods of strategising its presence in global affairs.

Implication: Government will become innovative and enterprising, enabling itself to compete with the business networks for market prestige and acclamation as well as to work in partnership with business on environmental issues.

8. A new global managerial elite

A new generation of international strategic managers is emerging, able to balance commercial with environmental and ethical pressures in producing corporate financial performance.

Implication: Able managers will switch among the three sectors, as well as work multinationally, in order to experience the different requirements of the changing marketplaces.

9. Regenerated urban regions

City regions provide economic opportunities and efficient land uses. Regenerated urban regions are essential in achieving global sustainability. This regeneration is providing a commercial bonanza for international business.

Implication: Cities will not only survive, but thrive as centres for cultural and economic development in the new century. Many of the most productive centres' innovation will be in less densely occupied 'edge cities', away from the traditional cosmopolitan centres. These will be regenerated as less dense clusters of garden districts characterised by low buildings and greater sensitivity to the human natural environment.

10. Environmental legislation and international organisations

There are now more than 25 global governmental and non-governmental organisations with a stake in business and the environment.

Implications: The growth of international organisations and global agencies with a stake in the environment has had profound implications for international business and for national governments in a rising number of conventions, codes, protocols and treaties which contain specific requirements for protection of the environment. Table 3.1 summarises some of the main ones.

It is only in the last half-century, since 1945, that such organisations with effective powers have become widely supported. In the second half of the century their presence has brought about a revolution in the way government is perceived. For example, by 1995 the Intergovernmental Panel on Climate Change, a group formed under the

Table 3.1 Some global and transnational agencies with a stake in the environment in the 1990s

	Intergovernmental and UN agencies	*Regional focus*	*Sectoral focus*
1	UN Environmental Programme UNEP	Developing countries, especially less developed	Environment
2	UN Education, Scientific and Cultural Organisation/ UNESCO	Developing countries, especially less developed	Education and social development
3	World Health Org./WHO	Global	Preventive and healthcare services
4	International Labour Organisation/ILO	Global	Employment conditions and structures
5	World Food Organisation/ WHO	Developing countries	Food supplies, markets and agriculture
6	International Monetary Fund/IMF	Global	Balance of payments and development loans, economic restructuring
7	Framework Convention on Climate Change	Global	Greenhouse gas production
8	Organisation for Economic Cooperation and Development/OECD	Developed countries	Economic and social infrastructure, comparisons and best-practice transfers among members, data
9	Convention for Sustainable Development/CSD	Global	Global environment
10	World Bank	Developing countries	Capital and human infrastructure development
11	World Meteorological Organisation/WMO	Global	Weather modelling and prediction
12	European Monetary Institute (to become the European Central Bank)	Members of the European Union	Taxation and monetary policies, institutions
13	United Nations Univ.	Global	Research
14	Intergovernmental Panel on Climate Change/IPPC	Global	Global warming
15	European Bank for Reconstruction and Development/EBRD	Eastern European countries	Economic and social redevelopment
16	International Climate Control Convention/ICCC	Global	Global warming
17	International Maritime Organisation/IMO	Global	Regulation of shipping and sea-based economic activity
18	UN Convention on Biological Diversity	Global	Biological research and prediction of population changes
19	UN Habitat	Global	Housing and urban planning

Source: BSRU U. of Greenwich 1996.

Table 3.2 Some major voluntary non-governmental organisations (NGOs) with a stake in the environment in the 1990s

Voluntary non-governmental organisations (NGOs)			
1	World Business Council for Sustainable Development	Global (international trading and manufacturing corporations)	Environmental impact and mitigation with responsible business practice
2	Global Climate Coalition/ GCC	Global	Industry-based lobby group; research on climate
3	International Chamber of Commerce	Global	Environmental management and sustainable businesses
4	Greenpeace	Global	Activist response to environmental disasters generated by government and corporate procedures
5	World Wildlife Fund/WWF	Global	Public awareness and lobbying for species protection
6	Friends of the Earth/FOE	Global	Public awareness and lobbying for environmental protection
7	Environmental Defence Fund/EDF	North America	Research, environmental issues
8	Worldwatch Institute	Global	Statistical research on global change, demography

Source: BSRU U. of Greenwich.

auspices of the United Nations in the aftermath of the Rio conference in 1992, formally announced at their world meeting in Rome that a link between human activity and climate change could be scientifically demonstrated. Over 2,500 scientists working in three groups across the globe were involved. Table 3.2 indicates some of the better known NGOs in the international sector.

Working Group I analysed the functioning of the climate system and potential changes to it resulting from human activities. Working Group II assessed potential impacts of climate change, adaptation strategies and measures that could be adopted to reduce greenhouse gas emissions. Working Group III focused on evaluating the economic implications of climate change. Secretary-General of the World Meteorological Organisation, Godwin Obasi, told delegates at the 1995 opening ceremony that the report was a 'warning to humanity that we have gone beyond the point where sustainable use of the atmosphere as a highly mobile dump for man's waste is possible without serious consequences'.[1]

CORPORATE RESPONSE TO THE TEN TRENDS

The rise of environmental commitment and the reporting cycle

In the midst of this process of change, international companies have been increasingly committing themselves to finding methods of building their business with environmental sustainability in mind. The formation of the World Business Council for Sustainable Development (WBCSD) in the mid-1990s was a major step in the direction of environmental responsibility and will be discussed in Chapter 4.

The insurance industry, badly damaged by environmental litigation during the 1980s, had, by the late 1990s, developed specific approaches to the environment. Twenty-two top companies, including Swiss Re, General Accident, Sumitomo Marine and Fire and National Provident Insurance, jointly signed a pledge to adopt best environmental management practices and to encourage their clients to do likewise. While American insurers were slow to join, every other major insurance group in the world was represented. It followed a similar grouping of commercial banks who made environmental commitments in their global associations.

In the insurance field this move has deep implications. The pledge essentially harnesses direct commercial interests of insurers in reducing risks and claims by companies and governments. Prevention of marine oil spills and reduction of greenhouse gases are seen by the insurance industry as an important focus in maintaining their viability. If catastrophic risk becomes more common, the insurance industry simply disappears. No one can afford the premiums. According to UNEP more than $50 billion in claims arose from just 12 storms between 1987 and 1995 – including Hurricane Andrew in Florida. Han Alders, UNEP European director, says that, 'Insurers know that environmental risks are business risks, pure and simple. They know that a few major disasters caused by extreme climate events or toxic waste spills could literally bankrupt the industry in the next decade.'

In spite of these signs of progress, in 1995 only 110 of the world's 37,000 multinational companies produced separate environmental reports. Among many commentators, John Elkington of the London company SustainAbility, pointed out the two major changes that drove up this number dramatically in the following years. The first was the adoption of global and European standards for environmental management, ISO 14001 and EMAS (the European Union's environmental management and audit scheme). Both offered a way forward for companies wishing to begin environmental reporting but without access to expensive consultancy assistance.

Secondly, and more crucially, the financial community – banks, insurers, stock markets, and intermediary institutions – began environmental risk rating in 1996. This procedure, long discussed, began to clarify which factors will be pivotal in pricing risk in corporate lending, investment and insuring. For the first time, key stakeholders were able to price potential liability in their investment in a company, as well as potential gain.

The method of identifying and pricing risk built on many existing conventions in accounting and investment practice. The use of Discounted Cash Flow, for instance, allows time-based calculation of risk, just as it does for profit. Later profits are worth less in the here and now, just as later risks are less dangerous than earlier ones from the point of view of the present moment of calculation.

However, the environmental-risk game for companies has proven in many cases a disaster area since 1980. The long projection of risk seldom covers the near-term likelihood of accident and managerial ineptitude. Increasingly the managerial and technological resource base of companies was seen as requiring major investment in order to cope with potential threats from environmental, safety and health impacts of business.

Training for environmental management

In a number of countries, companies' rapid acquisition of environmental reporting skills under the impetus of regulators and customer expectations propelled large numbers of managers from a variety of disciplines into the role. There emerged national training and certification institutions to oversee this quasi-profession. For instance, in the UK the Institute of Environmental Management (IEM) launched schemes to raise training standards. They reported on the current knowledge base of UK environmental managers. In a 1995 poll, and despite a generally sound grasp of technical and legal issues, many applicants for associate membership displayed no broad grasp of the strategic levels of sustainable development and what it might mean for their employers and clients. This level of professionalism would, of course, require many more years of programme and disciplinary development.

Organisation and the environment

Even while international organisations were expanding, the technology available through computers and telecommunications was changing the way all organisational problems were approached. This is especially true in government.[2] As in centralised administrations in the corporation, government bureaucracies are unable to cope. As a result of the information economy, there is an international trend away from centralised government and to locally driven autonomous government units. The consequence is that federal and international organisations are providing a second level of reference for local government seeking to expand its powers.

International and regional organisations provide a countervailing level of authority. They have become, particularly in federally structured regions like Canada and the European Union, a means by which local governments, citizens and small groups can seek redress against central governments. These loose supranational structures are coming to resemble not so much strong national states as the evolution of older and tighter federal networks such as those of Canada and the Russian Federation, once the core of the USSR. These national structures have been subjected to many decentralising stresses in recent years. Yet they retain a key element for their constituent parts: the option of moving into total independence as a small state threatens the loss of all sense of control as world power struggles are decided by deals among the largest nations and blocs of nations. During this era of decentralisation, ironically, weak international organisations see their roles expand as their capacities are used to balance the monoliths of large corporations and centralist national governments. Nowhere is this pattern more clear than in the case of environmental legislation.

In Europe, for example, the European Commission enacted the 1988 Directive on environmental assessment by which all 'major projects' – defined in the Directive – in member states were to be subjected to environmental assessment (EA). In some countries the EA directive was adopted consistently and it was expanded to include many kinds of projects. In others it has been virtually ignored. In 1994 the European Union's Environmental Management and Audit Scheme, a voluntary approach aimed at larger private businesses, was established. This approach in contrast to environmental assessment was voluntary and required both governments and companies to volunteer to participate. As an incentive, it offered international recognition to companies willing to bring their environmental performance into public view.

THE NEW COVENANT

As a result of changes identified in the ten major trends and the corporate response to those trends, a global triad of corporations, non-governmental organisations and governments has emerged. Government has led the environmental change process since the 1970s. Through regulation and penalties, it has provided the major classical force for environmental management. However, in recent years business has articulated itself as a second force in the environment. Key global enterprises are establishing and forcing a competitive advantage for themselves through environmental performance. Simultaneously, as an emerging form of 'feel good' service provider, Greenpeace and other NGOs of the third sector are challenging religion with secular allegiance to non-material values (see Chapter 5).

As a result of competitive and co-operative impulses among these three groups of players in international environmental diplomacy and law making, a *new covenant* has been achieved. This is so for the direct reason that environmental change requires a subtle evolution of capacities. Government must evolve co-operative methods to gain the commitment of corporations. In turn, corporations must commit to open disclosure and public reporting of their environmental performance, to establish trust and sustain long-term stakeholder confidence. And NGOs must design systematic planning, marketing and financial methods to be able to project their product – that is, how to save the environment – into countries which lack sufficient environmental controls to ensure the human future and to ensure they can survive as enterprises without depending on unpredictable waves of charitable donation. What are the implications of the new covenant for newly industrialised countries? For international corporations? For the future of the global triad?

COUNTERVAILING CASE NO. 1: NEWLY INDUSTRIALISED COUNTRIES

In the case of developing countries, environmental problems are inextricably entwined with development issues. The tendency of cash-strapped governments of developing countries has been to argue that either economic growth must come before scarce resources are expended to solve environmental pollution, or the wealthy states in the international community should provide financial and technical assistance to address the environmental issue. More recently, attention has been focused on the linkages between growth and environmental pollution, with dramatic illustrations. For example, a 1997 World Bank study estimated the cost to China of air and water pollution at $54 billion a year, equivalent to 8 per cent of its GDP.[3] Many developing governments are now adopting environmental regulations which are modelled on standards developed in Europe and North America, but the inability of these governments to enforce their regulations remains a serious obstacle to improvement. For example, in 1997, Ibama, Brazil's environment agency, collected only 6 per cent of the fines it levied for illegal logging in the Amazon.[4] Lack of uncontaminated water, raging forest fires, severe air pollution caused by soft-burning coal for heating homes and industrial emissions, mountains of untreated waste, squatter settlements and rapid population growth which places even more severe stress on the environment, combine to create both a sense of urgency for resolving the problems and a feeling of frustration at the enormity of the task. All three sectors of the global triad have been engaged in the question of how best to assist the developing countries, especially the poorer ones. Their concern will intensify as the problems of poverty in the third world continue to affect business, governance and quality of life in the developed world. In turn, as developing countries try to solve environmental and development problems through privatisation strategies, private firms will play an ever-increasing role, sometimes with mixed results.

Case: hydroelectric dams in Malaysia

In Bakuan Sarawak, the Malaysian government is promoting a giant hydroelectric dam to lead the development of what they regard as a backward region with low population. The loss of hundreds of protected species of animals and plants has been projected. Bakuan is the last considerable area of rain forest left in Sarawak.

The Malaysian government has until the late 1990s been praised by its business elites for its 'pragmatic' attitude towards conservation and green issues. On the whole, development has been unhindered by sustainability issues. But tourism is as important for growth as manufacturing and the agro-industry. Tourism industry leaders perceive the Bakuan project as 'over-developed'.

Defending the government, ministers allege that these critics are concerned that their reputation as a national tourism destination will be damaged by global allegations of inappropriate and unsustainable development. The government remain confident that Malaysia's tourism reputation will be conserved in other areas and that the Bakuan project will have many positive social and ecological benefits. As well, they argue that rainforest conservation is unrealistic and will, in the end, collapse under the weight of population and economic forces. Clearly deforestation can be attributed to a number of factors, not just the building of dams. Commercial logging, local demand for firewood, financing war with timber, slash-and-burn cultivation, need for cleared agricultural land and uncontrolled fires also take a heavy toll. Tourism can also be promoted on cultural and historical dimensions. The debate, however, will continue over whether dams are desirable, how much of the forests should be conserved and what the development costs of that conservation will be. Furthermore, with the economic retraction and market dislocation occurring in Southeast Asia in late 1998, the resources to continue developmental projects may well be restricted.

Case: newly industrialising countries and suburban golf courses

Urban growth and the new international commercial middle classes are profoundly changing the development of urban regions. Across East and South Asia, pressures on land use have come together in the problem of urbanisation, a problem clearly presented by a BBC4 programme in 1996. Conflict is embedded deeply between sustainable pre-industrial agricultural communities on the one hand and, on the other, international industrial development elites driving toward the economic advancement of their previously backward societies – as they see them.

Japanese companies have since the early 1980s been involved in a rapid overseas investment programme. They have built many productions facilities in Indonesia, Malaysia, Borneo, Australia, India, Vietnam, and the Philippines. Local business elites and associated government leaders have all benefited in a vast explosion of new wealth.

As the British and Americans before them, the Japanese have exported not only their industrial and commercial skills, but also their lifestyles and preferred leisure pastimes. Number one among these is a passion for golf which to Western audiences is somewhat incomprehensible. It is difficult to understand because it appears to affect almost every element of ambitious male corporate culture, whereas, popular as it is in the West, it still attracts only a fraction of adult men and women.

In the rapidly industrialising settings of the Pacific Rim, golf is the great non-verbal shared passion which can be used to knit together deals which otherwise might not gel. It overcomes the problems of language, culture, custom, in a physical sport which can be played by men of any age. Few Asian women participate in golf outside Japan. The question to be raised is as follows: Does the presence of a golf course in a developing country represent good ecology? Good social development?

In 1996 around the capital of Jakarta, some 25 new golf courses had been created out of farmland over the previous decade. As many as 50 similar golf projects are proposed in the urban fringe of this metropolis of 15 million people, although the financial and economic difficulties of Japan and the political instability of Indonesia in 1998 may curtail these projects.

Local farmers say that without compensation they have been cheated out of and forced from their tribal lands. Such lands have been farmed at ecological and sustainable levels for hundreds of years. Land ownership in such societies has been in many cases through local community and tribal councils. It is these groups that have struck deals with golf-course developers.

The typical joining fee for a business leader joining a golf course is US $70,000. Annual fees plus meals and other subsidiary incomes may approach additional tens of thousands. The appetite for golf as a major focus for senior business dealing is so great that these giant green reservations, preserved for a relatively tiny minority, can be likened to stock exchange trading floors of a previous generation. In such vast reservations, surrounded by a teeming, impoverished population of excluded farming communities, multi-billion-yen relationships can be cemented.

To lubricate this process, a full panoply of food, alcoholic beverages and other leisure drugs and, it is alleged, prostitutes, are made available. On one golf course, the manager notes, 'All our caddies are female, unmarried and under the age of 27. When they marry, they are expected to leave us. By local standards, they are well paid.' He then notes that the tape recorder is still running: 'I shouldn't probably be telling you this; you won't quote me, will you?'[5]

Responses to counter the trend toward more and more golf courses has only recently been evident. In Japan a young woman lawyer, fluent in English after many years in the USA, formed in 1993 the Global Organisation Against Golf Courses (GOGAS). This group has found instant world-wide support. The anti-golf movement in the developing world is only emerging in the late 1990s but forms an important reassertion of rights to land use for deprived majorities of urban dwellers.

To these critics, the culture of the golf course appears to bring together all the worst manifestations of colonialism, imperialism, misogny, and economic and cultural exploitation. It destroys the well-being of traditional farmers, communities and cultures and subordinates them to the deal-making culture of international economic elites. Crowded tropical countries with dense populations struggle to maintain a subsistence agrarian economy, even as they develop. In time, the proliferation of golf courses around large cities may be perceived as an environmental crime ranking with the closure of the English common lands of the nineteenth century and the loss of Amerindian lands during the eighteenth, nineteenth and twentieth centuries. The persistence of destitution and poverty in developing countries continues to drive change in the direction of universal environmental and ethical regulation by governments. World business, increasingly faced with a common global standard of moral expectation from community opinion and the voluntary sector, must respond. Whether that response will be to pursue 'business as usual' strategies or to act innovatively in addressing the economic and social impact of overseas investment remains to be seen. The second countervailing case illustrates the choice which is open to international firms.

COUNTERVAILING CASE NO. 2: SPLITTING THE ENVIRONMENTAL CONSENSUS AMONG LARGE INTERNATIONAL CORPORATIONS

Case: the global climate coalition

The complexity of relationships among government, NGOs and business began to be apparent as corporations lined up to focus on environmental treaties and global warming.

By 1996 the forces arguing for major industrial change were hardening into two opposing camps. For ten years, general uncertainty of data analysis in the scientific community led to the question: Was global climatic change really occurring or were recent climatic variations simply natural perturbations in the weather cycle of the biosphere? In 1996 a group of 5,000 scientists, drawn from many different countries, were able to agree that change triggered by human society existed and that the direction was toward global warming. These findings were published world-wide just prior to the UN Climate Change Convention meetings in Geneva in the summer of that year.

During this period many businesses, including large international firms, had found it prudent to begin to assume that change was under way. They identified with the Rio Business Sustainability group of 1991 organised by Stephan Schmidheiny of Switzerland, a top corporate executive, and author of the world best-seller, *Changing Course*. (See Chapter 4.) Other firms, especially oil and other energy industries, took a more confrontational view. Global warming, they said, may exist; but it would take 50 years to prove it. Their view was that variations of climate in the 1980s and 1990s were replicated in historical weather records. The amount of energy and pollution coming from combustion of fossil fuels could not make that much difference in a biosphere which, after all, had endured volcanic eruptions and meteorite showers as substantial contributors to its history of change.

By the middle of the twenty-first century, these companies asserted, there would be alternative energy technologies produced by the workings of the marketplace. At that time, if there were still reasons for changing to non-fossil fuel energy production, there would be minimal disruption to the development of the world economy. All this was argued within their own non-governmental organisation, the Global Climate Coalition (GCC). The GCC, an organisation of mostly US private firms and business trade associations with headquarters in Washington, D.C. was formed in 1989 'to coordinate business participation in the scientific and policy debate on the global climate change issue'.[6] GCC sponsors research on earth systems and climate models, as well as the decision-making process which leads to solutions for addressing climate change. It was visible at the Rio meetings in 1992 and in subsequent UN meetings at Kyoto in 1997, particularly as it challenged the assumptions and data of green groups. Its membership, numbering 65 as of 21 May 1998, is presented in Table 3.3.

Table 3.3 The global climate coalition: members in 1998

Board membership	*Board membership*
ABB Vetco Gray, Inc.	Air Transport Association
Allegheny Power	Ameren Services
American Automobile Manufacturers Assoc.	American Commercial Barges Line Co.
American Farm Bureau Federation	American Forest and Paper Assoc.
American Iron and Steel Institute	American Petroleum Institute
American Portland Cement Alliance	Amoco Corporation
American Farm Bureau Federation	Atlantic Richfield Coal Co.
American Iron & Steel Institute	Chemical Manufacturers Association
Chevron	Chrysler Corporation
Cinergy	Council of Industrial Boiler Owners
CSX Transportation, Inc.	Cyprus-Amax
Dresser Industries	Drummon Co.
Duke Power Co.	Edison Electric Institute
Exxon	Ford Motor Co.
General Motors	Illinois Power Co.
IPSCO Steel Inc.	Mobil Corporation
National Association of Mfgs	National Lime Association
National Mining Association	National Rural Electric Cooperative Association
Norfolk Southern	Parker Drilling Co.
Petroleum Helicopters, Inc.	Process Gas Consumers
Rail Progress Institute	Sante Fe International Corp.
Southern Company	TECP Energy, Inc.
Texaco, Inc.	TRICO Marine Operators, Inc.
US Chamber of Commerce	Virginia Power
Western Power	
General membership	*General membership*
BHP Minerals	Bethlehem Steel
Consumers Energy	ELCON
Eastman Chemical	Goodyear Tire & Rubber
Greencool	Hoechst Celanese Chemical Group
Kaiser Aluminum & Chemical Corp.	McDonnell Douglas
Northern Indiana Public Service Co.	Ohio Edison (power)
Society of the Plastics Industry	Union Carbide
	Union Pacific

As should be obvious from the membership list, the Global Climate Coalition is essentially an association of American firms and it is 'global' only in the sense that it has primarily addressed the issue of global climate change. Its claims are indicative of the concerns and values of this lobby group as the following quotation illustrates: 'Large manufacturers in the iron and steel and paper industry join small businesses with common interests in maintaining the abundant and inexpensive energy that keep American standards of living the envy of the world.'[7]

The view of the critics

From the perspective of the many 'green' organisations, of course, this NGO was a source of great irritation. Although it wore the same clothing as the other global change NGOs, it was, in fact, a lobbying organisation heavily funded by some of the most powerful corporations on the planet, including the largest fossil-fuel producers.

According to its critics, its fundamental hidden agenda was to scupper pressures to change governmental policies world-wide away from fossil fuels and towards sustainable sources. It was seen by 'greens' as a crass trade-protection body, cleverly disguised as a *pro bono* (for the public good) voluntary association. An environmental journalist and commentator, Paul Brown, notes:

> For those who have attended the dozens of scientific meetings of the Intergovernmental Panel on Climate Change (IPCC), and the parallel political gatherings, the fossil fuel lobby have been a constant burden. As scientists and politicians try to agree internationally on addressing the greenhouse problem, there are men in suits making objections – they interrupt every meeting, stalk delegates in coffee bars and corridors, all to delay, obscure and if possible stop action. Their perseverance and dedication, backed by unlimited resources, show the importance of their mission to preserve the short-term interests of the fossil fuel lobby at all costs.[8]

To many observers, GCC's support for scientific research is nothing more than an affirmation of the wisdom of cheap energy in the American economy, with the consequences for excess atmospheric carbon production this brings about. From this perspective, the continued presence of GCC in the international policy and scientific meetings on global warming suggests deliberate manipulation by powerful and well-funded interests.

Among the early membership of GCC were such firms as BP and Shell, both of which claimed to be environmentally aware petroleum companies; indeed BP owned its own solar energy subsidiary. Some UK shareholders argued that BP should not associate with this industrially-based NGO if its environmental claims were in any way consistent with its actions. Both Shell and BP have subsequently withdrawn and joined the World Business Council for Sustainable Development. But lobbyists argue that business does have a right and duty to defend the interests of both their shareholders and other stakeholders. Certainly neither scientists nor politicians can claim to own proprietary rights to any central set of 'truths'. Major businesses cannot be excluded from attempting to influence the direction of international environmental policy solely because they are businesses. And, in fact, all business-oriented international organisations, including the International Chamber of Commerce, the World Business Council for Sustainable Development and the GCC, see that as their mission. The difference among various international business organisations must be established on their track record in following environmental best practices, however incrementally, and not on their fundamental purpose nor on their country of origin. In an era where intelligent consumers and ethically-driven legislators are common, however, efforts to distort the policy-forming process appear ultimately damaging. The associate companies and industries increasingly will be shown to be self-seeking on an issue of over-arching public importance.

The insurance industry breaks ranks

A more impressive development in the late 1990s was the split in the global business community itself. The world insurance industry had carried out its own meteorological research and had publicly accepted that the number and force of major weather incidents – storms, hurricanes, tornadoes – were increasing to levels that were unprecedented. Frank Nutter, president of the Reinsurance Association of America, warned the US Senate in 1996 that climate change might make the industry insolvent. His association reported that claims from weather-related disasters rose from $16 billion for all the 1980s to $48 billion for just the first half of the 1990s.[9] In other words the rate of claims in the 1990s was approaching six times those of the 1980s. As more and more firms began to recognise the potential costs of environmental change and destruction, as well as the impact of increasing governmental regulation, a significant shift in attitude toward the challenge of sustainable development occurred. By 1998, approximately 80 international treaties were being presented to national governments for agreement at the rate of 8 or more per month. Governments, not-for-profit green organisations and lobbyists were all swamped with the velocity and complexity of the agreements being considered. The response of international firms to growing international environmental regulation is the subject of the next chapter.

4 Environmental Diplomacy and Corporations in the World-Wide Web of Interaction

Reality Field 4
Multinational Corporations, as part of the global triad, are actively shaping international environmental treaties and working in collaboration with governments and non-governmental organisations to develop global best practices.

INTRODUCTION

Earlier chapters have indicated the range and variety of conflicts arising among commercial development, community social life and the natural environment. All of these experiences can be grouped under the much-criticised trend of globalisation. Methods of Western corporations are adopted in developing countries even where local businesses drive change. A critical function of interlinking international standards, business, social and environmental development is the discrete set of institutions and processes summarised in the phrase 'environmental diplomacy'.

Diplomacy, the conduct of sovereign states in reaching agreements among them is customarily treated as a state function carried out solely by official representatives. Their activities are guided by an over-arching foreign policy. Using a more expansive concept of diplomacy, this chapter presents a different perspective on sovereignty and on the process of environmental diplomacy, which is bringing about hundreds of new agreements and treaties. It encompasses the growing influence and interaction of unofficial groups in the formulation of international environmental agreements, namely multinational corporations (MNCs) acting as non-governmental organisations (NGOs).

Searching for agreement on international rules of conduct inevitably involves discussion, negotiation, tradeoffs and pursuit of interest, whether collective or individual. Consequently, interaction among interested groups and states in international and regional organisations creates a complex, multi-layered and multi-dimensional, world-wide web of interaction and influence. This web is sufficiently dense to prevent any single group or state from exclusive control of outcomes, particularly in the context of negotiating environmental agreements in international organisations such as the United Nations.

Furthermore, what induces co-operation, tradeoffs and compromise among contending groups is precisely this density of influence-seeking interaction. This occurs regardless of how disparate their views may seem and regardless of how dominant one group or state may appear to be.

The role of governments and NGOs has received considerable attention in the negotiation of environmental treaties, but little analysis has been focused on corporations' action internationally in a role which is remarkably similar to that of environmental non-governmental organisations in the not-for- profit sector. Especially in the context of negotiating international agreements on environmental issues, multinational corporations not only attempt to influence policy outcomes, use knowledge as power and offer tradeoffs. They also demonstrate an increasingly strong commitment to environmental goals. In order to understand the role firms play in international negotiation and the implication this role has both for international relations and for diplomatic practice, we will examine the role of multinational corporations in the United Nations, in a regional organisation and *vis-à-vis* other non-governmental organisations.

AN EXPANDED CONCEPT OF DIPLOMACY: WHO PARTICIPATES?

The phrase 'environmental diplomacy' is now widely used to describe international negotiations which result from a desire to address the problems of environmental degradation and pollution on a global basis. Because of the complex and interacting nature of environmental issues, the number and variety of groups involved in the process of writing and agreeing global conventions has increased exponentially. This increase, in turn, has raised the question of who should be involved in global negotiations and on what basis should participation be established. To answer the question 'Who should be involved?' requires firstly an assessment of the nature of diplomacy itself. Does its origin or its historical development provide any insights to the fundamental question of participation?

Even a cursory survey of historical writings on the origins of diplomacy reveals a continuously shifting focus as well as definitions which are specific both to time and to context.[1] Use of the word 'diplomacy' as well as the contemporary practice of diplomacy has been characterised by a strong emphasis on the exclusive role of state officials, and the task of defining diplomacy tends to be cast in terms of official representation.[2] Simply put, the idea is that only diplomats, because of their specialised training and experience, and because they represent sovereign states in a state-centric international system of states, should negotiate international agreements, preferably in closed sessions with fewer participants. In this setting, it is argued, professional diplomats will be able to bring the art and science of diplomacy to fruitful conclusion. This classical view of diplomacy is strongly reinforced by the classical international legal perspective which emphasises the contractual nature of treaties, promulgated by states alone, and which play a central role in formulating the principles of international law. If the classical view were an accurate reflection of reality, non-state groups such as multinational corporations would have little scope for action.

Diplomatic negotiations: exclusion or expansion?

The emergence of democratic states in Europe during the eighteenth and nineteenth centuries and the rise of new states in the twentieth century gave impetus to an increasing tendency toward large numbers of non-diplomatic advisors attached to diplomatic delegations attending international conferences. In turn, the assessment of what constitutes successful diplomatic practice tended to reflect a fundamental division between those who advocated a return to the classic tradition of exclusivity and those who emphasised the more expansive practice of diplomacy in the nineteenth and twentieth centuries.

More recently, exclusionists and expansionists, struggling to encompass the increasing complexity of international negotiation and the role of diplomats in that process, have been driven to ever greater creativity and inventiveness in their definitions of diplomacy. In one widely quoted work alone, there are references to twelve different modifications of diplomacy: open and secret, moral, old and new, boudoir, democratic, small power and great power, oriental, conference, and amateur diplomacy.[3] In practice diplomacy has never been anchored to a fixed and unchanging definition. Its strength lies precisely in its flexibility.

Variable definitions of diplomacy

A sampling of the principal definitions of diplomacy offered by diplomats and diplomatic historians over a period of three centuries provides evidence of the variable use of diplomacy as a concept:

- diplomacy as a profession, 1716[4]
- the science of external relations, 1811[5]
- the knowledge of the principles, maxims, skills and proprieties necessary for the external management of public affairs and businesses of the people or the sovereign, and for verbal and written negotiations with foreign states, 1818[6]
- gathering the knowledge and the principles which are necessary to conduct public affairs effectively between states, 1846[7]
- the science or art of negotiation, 1866[8]
- the art of conciliation of the interests of the people among themselves, 1883[9]
- the application of intelligence and tact to the conduct of official relations between the governments of independent states, extending also to their relations with vassal states, 1932[10]
- the management of international relations by negotiation; the method by which these relations are adjusted and managed by ambassadors and envoys; the business or art of the diplomatist, 1939[11]
- maintenance of peaceable relations between tribes and nations, 1950[12]
- mediation of the mutually estranged relations of states; mediation of estranged peoples organised in states which interact in a system, 1987[13]
- the conduct of international relations by negotiation rather than by force, propaganda, or recourse to law, and by other peaceful means (such as gathering information or engendering goodwill) which are either directly or indirectly designed to promote negotiation; it is a professional activity, though non- professionals now play an important part in it, 1995.[14]

The last definition successfully captures the expanding milieu of late-twentieth-century diplomacy, but only J. Schmelzing's definition in 1818 reflects the long history of trading firms influencing negotiation. Nor would it have encompassed the role played by earlier multinational corporations such as the East India Company, the Hudson Bay Company in Canada or the Kimberly Mining Company and De Beers in South Africa. Even today, analysis and discussion of diplomacy neglects the role of firms, whether in nineteenth-century arms control negotiations or in twentieth-century environmental conventions.

Whilst it certainly remains the case that only officials appointed to negotiate on behalf of states can *sign* international treaties, the drafting, negotiation and implementation of those agreements not only depends on the actions of many other states. It also depends on the support of several national ministries and departments, on independent experts or specialists, on academics, on significant interest groups acting internationally, on parliaments and assemblies and on public opinion. To insist that only official representatives of sovereign states can negotiate treaties is to adopt an overly-constricted understanding of the dynamic and complex international policy process which has produced UN conventions on the environment. In particular, it does not encompass the influence that NGOs, both environmental and industrial, exert in shaping the agenda of those negotiations and in monitoring compliance with agreed rules.

Precisely because governments increasingly rely on non-state groups to help address and resolve international problems posed by the demand for sustainable development and by the desire to limit further damage to the environment, the practice of diplomacy has evolved in an expansive direction. That expansion gives non-governmental organisations, including industrial ones, a larger role in determining the content of treaties and conventions. That being the case, environmental diplomacy should not be conceived in its narrowest sense of states approving treaties that deal with environmental issues, but rather as a very broad, all-encompassing process which depends upon a myriad of non-state groups for its success. Demands for reform and criticism from active interest groups, specialists, the media and multinational corporations not only influence government decisions, but those same groups are essential for implementation and enforcement.

Sovereignty in practice: absolute or relative?

It is widely accepted that the classical tradition of diplomacy began with the study of states and international law. It is also accepted that this development of the state itself was accompanied by the gradual assertion of state authority, including the right to conduct warfare, to engage in diplomacy, to make treaties and to create public law. Developing simultaneously was a tradition of both international and internal regulation of abusive practices. Control and restrictions on the abuse of authority by princes and monarchs, both secular and religious, are reflected in just war doctrines, the treatment of prisoners, the establishment of truces, the immunity of diplomatic agents, the sanctity of agreements and the prohibition of noxious weapons.[15] These restrictions, in mediaeval and modern Europe as well as subsequently in those areas colonised during the epoch of European imperialism, constitute the basic foundation for the establishment of normative ideals, which were both practical and necessary for the wellbeing of the state, its subjects and its citizens. Hence the history of the modern state is a history not only of control and domination by a centralising authority claiming sovereignty, but also a history of reciprocal and accepted restrictions on the excess of that authority which in turn limited sovereignty.

In the international arena, the addition of new states, both within Europe and beyond, described in the previous section, also contributed to a contextual limitation of sovereignty induced by demands for increased participation and by the need to accommodate broader interests. Although traditionalists continue to assert that sovereignty is indivisible and absolute, the history of state practice would suggest otherwise.[16] Absolutist versions of sovereignty fail to reflect the actual conditions of international co-operation under which the newly emerging states of Africa, Latin America, and the Caribbean were incorporated into either the system of states or the international community.[17]

They also fail to capture the essence of 'pooled sovereignty' in the regional institutions of Europe and to account for those instances in which international and regional concertation has facilitated the transition from abusive domestic regimes to forms more compatible with the ideal of a just state. Treating sovereignty as variable rather than as a permanent and fixed principle, allows a more comprehensive understanding of why sovereignty is sometimes evoked and sometimes not. Variable evocation is especially the case in situations where absolute control of decisions is replaced by shared and agreed policy-making under conditions of broad consultation with other states and with all interested parties. Because of their size, wealth and 'global reach', multinational corporations have been accused of weakening state sovereignty. Given the variable practice of sovereignty, it is a difficult and problematic assertion.

If it is the case that governments sometimes ignore the notion of sovereignty in instances where co-operation and/or help from others is more likely to serve national interests than self-reliant strategies, then what is the implication for how the conduct of international relations is conceived? It suggests the need for a perspective that accounts for co-operation as well as conflict in relations among states and that also accounts for the involvement of non-state groups in international negotiations.[18] What, then, is the case for analysing multinational corporations as non-governmental organisations and for viewing them as part of the solution to environmental problems?

CORPORATIONS AND THE CHALLENGE OF ENVIRONMENTAL DIPLOMACY

As of 1998, there were over 100 international and regional treaties and conventions on environmental issues, either being negotiated or in force. As the sampling in Table 4.1 indicates, these international agreements contain provisions which have profound implications for corporations, representing both opportunities or costs, depending on the attitude and policies of corporate managers and the provisions of each convention.

Not only has the number of conventions increased dramatically, but the range of issues covered by them has also substantially increased since the 1940s. Table 4.2 presents a summary of that expansion.[19]

As the above survey demonstrates, environmental issues are not a new development in world affairs, nor is corporate involvement in these issues at the global level. Furthermore, corporate interest in environmental policies and regulations is not a fashionable trend which will be here today and gone tomorrow. What has changed in the international arena, however, is the interlocking density of global arrangements, the increased demand for effective enforcement and a growing recognition that the global efforts to control environmental damage in the 1970s did not appreciably reduce environmental degradation. Both regional and national environmental policies have had significantly better results, in many instances as a result of government, corporate and NGO collaboration. As governments and committed environmentalists searched for solutions and experimented with environmental models, a growing awareness of the cost factor

Table 4.1 International agreements that affect corporate interests

International agreements	*Regional agreements and/or projects*
1. Convention for the Regulation of Whaling, 1946, 1931 2. International Convention for Prevention of Pollution from Ships, 1973 3. International Convention for the Prevention of Pollution at Sea, 1954, 1973 4. Convention on the Prevention of Marine Pollution by Dumping of Wastes and Other Matter, 1975 5. Convention on International Trade in Endangered Species, 1975 6. Convention Concerning the Protection of the World Cultural and Natural Heritage, 1975 7. Convention on Wetlands of International Importance, Especially as Waterfowl Habitat, 1975 8. Convention on Conservation of Migratory Species and Wild Animals 9. International Convention on Long-Range Transboundary Air Pollution, 1979 10. Convention on Conservation of Antarctic Marine Living Resources, 1982 11. Convention on Biological Diversity, 1992 12. International Coral Reef Initiative Treaty on Fishing, 1995 13. Kyoto Protocol to the UN Framework Convention on Climate Change, 1998 (not yet in force)	1. NAFTA, Commission on Environmental Cooperation (Canada, US, Mexico) 2. US–Mexico Air Quality Improvement District for El Paso, Texas and Cuidad Juarez, Mexico 3. Regional Environmental Centre for Central and Eastern Europe 4. Summit on Sustainable Development, Western Hemisphere, Santa Cruz, Bolivia, 1994 5. Six regional environmental hubs established by the US in Central America and the Caribbean, Costa Rica (loss of forest and biological diversity); Central Asian in Tashkent (water management in the Aral Sea Basin); Eastern Africa in Addis Ababa (desertification, deforestation, biodiversity loss and water use); South Asia in Kathmandu (alternative energy, clearn air, water sharing and environmental disaster preparedness); Middle East in Amman (water resources, desertification and coral reefs in the Gulf of Aqaba as part of the Middle East peace process); Southeast Asia in Bangkok (sustainable management of foreign and marine resources), 1997, 1998

gave impulse to the principle that 'the polluter pays'.[20] Both governments and corporations were threatened with a massive increase in costs, as well as an opportunity for benefits which created a

Table 4.2 Summary of environmental issues addressed in international conventions

Date	*Issues*
1940–1949	Natural resources, science and resources, whaling.
1950–1959	Sea pollution, population growth, living and non-living sea resources, atomic energy, fishing on the high seas.
1960–1969	Energy sources, national parks, nuclear weapons, heritage monuments and sites, outer space, world weather, science and the biosphere, natural resources, human environment.
1970–1979	Marine pollution, environmental future, national parks, ship pollution, environmental science, population growth, resource and development, waste disposal, endangered species, cultural and natural heritage, wetlands, environmental education, human settlements, water supply, desertification, primary health care, environmental modification, world climate, migratory animals, linking resources, population and development, science and technology for development, air pollution.
1980–1989	Antarctic marine (living) resources, renewable energy, plant genetic resources, renewable energy, water supply, sanitation, the human environment, whaling, outer space, national parks, environmental resource management, sea pollution, tropical timber, environment and education, biosphere, technical assistance for environmental clean-up, acid rain.
1990–1998	Climate change, ozone pollution, environmental monitoring and risk assessment, penalties and sanctions for pollution, relative costs and benefits, impact on global trading arrangements in the WTO, environmental conditionality, sustainability.

Sources: See note 19.

convergence of interests. The result of this convergence led to strategic alliances, in which the support of environmental groups was increasingly sought by both; hence our argument that a global triad, composed of states, corporations and environmental groups based on both goal-inspired and threat-induced co-operation is emerging.

CORPORATIONS AND THE UNITED NATIONS

Corporate interest in UN environmental activities is often said to have begun during the UN Conference on Environment and Development

(UNCED) in Rio in 1992 when the Secretary-General of the conference encouraged the participation of industrial groups.[21] However, just as the role of NGOs have evolved, companies, both private and state-owned, have also participated in international organisations since the nineteenth century and some of these activities were aimed at conservation issues. For example, Lynton Caldwell cites the example of an association of German farmers and foresters in 1868 who wrote a declaration addressed to the Foreign Office of Austria-Hungary, urging protection of birds and animals useful to agriculture and forestry.[22] Indeed, one of the earliest non-governmental organisations was the International Chamber of Commerce (ICC) formed in 1919, which had three places on the League of Nations' Economic Consultative Committee established in 1927 where it had full voting rights and participated directly in League negotiations.[23] It was subsequently given consultative status as a non-governmental organisation with the UN Economic and Social Council in 1946.[24]

Today ICC represents 7,000 companies and businesses in 130 countries and it continues to work closely with the United Nations through the Economic and Social Council (ECOSOC). In response to the pending Rio meeting in 1992, ICC sponsored a World Industry Conference on Environmental Management and produced a Business Charter for Sustainable Development, which was adopted by the ICC Executive Board in 1990. The Charter gave early indication that corporations would become increasingly involved in international negotiation of environmental issues.[25]

CASE: THE WORLD BUSINESS COUNCIL FOR SUSTAINABLE DEVELOPMENT

In a more pro-active stance, two European-based groups, the World Industry Council for the Environment (formed under the auspices of the ICC in Paris and the Business Council for Sustainable Development in Geneva) participated in the UN Conference on Environment and Development (UNCED) at Rio in 1992.[26] These two bodies subsequently merged in January 1995 to form the World Business Council for Sustainable Development (WBCSD). Both ICC and WBCSD have been active in international affairs, but ICC's charter emphasises its role as a defender of business interests whereas that of the World Business Council concentrates specifically on sustainable development and environmentally desirable business practices, in addition to its obvious interest in economic viability. WBCSD has been remarkably active and involved in global environmental issues since 1995 and it is this council which appears to have become a *de facto* non-governmental organisation. Although an ICC representative sits on its executive committee, WBCSD takes an independent stance and it is increasingly active in UN plenary sessions, *ad hoc* group meetings and conferences. There is, however, substantial overlap in the corporation membership of the two organisations.

The World Business Council for Sustainable Development: a corporate NGO

As of March 1998, the WBCSD represented 120 international companies in twenty major industrial sectors and located in 33 countries (twelve in Western Europe, four in Central and Eastern Europe, two in Africa, six in Asia, two in Oceania and two in North America). Current publications treat a wide range of environmental issues including eco-efficiency leadership, environmental assessment, cleaner production, biodiversity and business, by-product synergy and sustainable consumption, environmental performance and shareholder value. It has a statutory commitment to the 'environment, principles of economic growth and sustainable development' and its mission, articulated in the Annual Review of 1996, stipulates a mandate to

> provide business leadership as a catalyst for change towards sustainable development and to promote the attainment of eco-efficiency through high standards of environmental and resource management in business.[27]

WBCSD has a global network of 15 national, and one or two regional, business councils for sustainable development and four partner organisations, which represent approximately 600 members.[28] Regional councils for Latin America (which includes the USA) and for the Gulf of Mexico combine with the national councils to form a worldwide network for distribution of a quarterly newsletter, reports, books and briefing papers produced by various working groups and staff research. The new Foundation for Business and Sustainable Development is jointly funded by the EU and the Norwegian government and also relies on the UN Environmental Programme (UNEP) to implement its initial training programme.

The regional councils undertake a wide range of projects, mostly in developing countries, such as:

- helping local businesses in Columbia and Argentina achieve standards approved by the International Standards Office, 14000 series;
- natural resource planning in Costa Rica and El Salvador; shared workshops on eco-efficiency in Egypt;
- developing education programmes in Croatia, Zimbabwe and Russia;
- recycling waste in the Philippines.

Other projects focus on energy, trade and the environment, financial markets and natural resources. Regional and national councils are also the primary mechanism through which firms based in developing countries gain access to knowledge of the best practices for environmental protection. Current policy development has focused on five main topics: sustainable production and consumption, climate and energy, trade and the environment, financial markets, and natural resources.

During 1995–7, the WBCSD worked jointly with the World Bank, several UN agencies, the OECD, the World Trade Organisation and NGOs such as the Sierra Club, the World Wide Fund for Nature, the World Conservation Union (IUCN), and the Forest Stewardship Council. It has also cooperated with research institutes such as the World Resources Institute and the International Institute for Environment and Development in London. Major support and sponsorship, as well as personnel and research data, come from the member companies, over half of which are located in Europe and many of whom are major European-based multinationals. A sampling follows: Rhone-Poulenc, Volkswagen, Fiat, Pirelli, Philips, Unilever, ABB, Nestle, British Gas, BP, Glaxo Wellcome, Imperial Chemicals, Rio Tinto Corporation, Shell International, Zurich Insurance Group. The roster of American corporations includes AT&T, The Dow Chemical Company, Dupont, Eastman Kodak, General Motors, Procter & Gamble, Texaco, 3M Company and Xerox Corporation, and Japanese membership includes Mitsubishi, NEC, Sony, Toshiba, Nippon T & T and the Tokyo Electric Power Co.

WBCSD: innovative partnerships

In an effort to document the progress made by the Council, a special task force was formed to summarise the activities of WBCSD, the details of which were drawn from member corporations. The resulting report, *Signals of Change*, not only provides concrete examples of the 'responsible company' drawn from the experiences of 24 member companies and one of its regional councils with respect to environmental performance. It also indicates the extent to which WBCSD is linked with governments, international organisations and NGOs in supporting sustainable development.

A distinction needs to be made between the innovations which are directed at internal corporate governance in a single, nationally-based corporation and ones which are aimed at fostering sustainable development in developing countries. But in the case of multinational corporations with branch operations in developing countries, changes in internal governance that give environmental protection a higher priority are significant. Furthermore, the internal and national experiences of these corporations are the base for responding to UN environmental programmes, particularly those which involve financial assistance, the development of 'best practice', and technology transfers to the developing countries.

As the following examples illustrate, the number of programmes aimed at improving the environmental performance of WBCSD members is not only increasing, but is also addressing an ever-wider range of problems and solutions:

- 3 M's Pollution Prevention Pays programme, begun in 1975
- Product life-cycle environmental assessment at Xerox
- Environmental Performance Rating Units at Swiss Bank Corporation; Sony's 'Green Plus' programme
- Supply chain auditing now adopted by Fiat, B&Q, and IBM
- Closed-cycle paper bleaching in 15 paper mills worldwide (Sweden, Canada, Finland, the US and South Africa)
- Contribution of Ebara Corporation (Japan) to the UN University's Zero Emissions Research Initiative
- PowerGen's (UK) co-operation with the European Union in carrying out a pilot audit as part of EU's Eco-Management and Audit Scheme (EMAS)
- Dow Chemical's commitment to voluntary compliance programmes in the US, Canada, Europe and Australia
- A tree plantation policy in Aracruz Celulose (Brazil) which emphasises replanting that preserves biodiversity
- General Motor's sponsorship of Chinese research and development projects in electric-powered vehicles, technology seminars and educational/training programmes.[29]

Some of these initiatives have been criticised for being purely public relations strategies, for using unverified information, for permitting free rides, for being voluntary rather than compulsory, for failing to provide for third-party verification and for not necessarily improving performance. What is surprising, however, is that these criticisms are openly acknowledged and directly addressed in the Council's publications.

MNCs and climate change

The International Panel for Climate Change, established jointly by the World Meteorological Organisation and UNEP in 1988, is responsible for developing the 1992 UN Framework Convention on Climate Change. It provides a clear case-study of corporations responding to action by the UN by attempting to influence both the agenda and the method of assessing outcomes. ICC and WBCSD both have observer status, along with over 100 groups, at the Conference of the Parties; as such they are invited to participate in *ad hoc* meetings as well as plenary sessions of the Conference of the Parties.[30]

The Framework Convention and its subsequent Kyoto Protocol negotiated in 1997 and opened for signature in March 1998, contain national commitments to reduce six greenhouse gases which were not included in the 1987 Montreal Protocol on Substances That Deplete the Ozone Layer. Annex A of the Kyoto Protocol makes specific reference to four sectors associated with the production of these gases: energy, industrial processes, agriculture and waste. Both ICC and the World Business Council attended meetings and issued statements throughout 1996 and 1997 which were directed toward influencing the provisions of the Climate Change Convention. Corporations, acting through non-governmental organisations such as ICC and WBSCD, are clearly responding to pending regulations in both national and international law which will have a direct impact on their profits, on their stakeholders and on their public image.

As the intensity of preparations for the Climate Change Convention in Kyoto in December 1997 increased, business and industry associations were successful in adding new elements to the international environmental agenda which were of particular interest to them:

- unexamined effects of successful re-cycling of energy and waste on scientific projections
- international competitiveness

- the role of joint ventures in the development strategies of developing countries
- job creation
- common standards for industry
- industrial ecology
- the need for consistent and standardised monitoring of emissions
- incentives for business
- the possible conflict between national environmental standards and international free trade
- land use
- ecosystem management
- market opportunities available to corporations and companies who pursue a 'green' strategy.

Although the Kyoto Protocol was opened for signature on 19 March 1999, it will not come into force until at least 55 nations have ratified it and that ratification must include nations which accounted for at least 55 per cent of the total carbon dioxide emissions for 1990. Furthermore, Article 3 of the Protocol gives the Parties until the year 2012 to reduce their overall emissions by 5 per cent below the levels existing in 1990. Whilst this lengthy approval and implementation period may appear to deny the urgency of global warming and has been deplored by some NGOs, it reflects the successful attempt by corporate NGOs to modify scientific research data and to provide additional time for industry to comply with its provisions.

Increased corporate concern was also apparent in May 1996 when the UN Environmental Programme (UNEP) sponsored a conference entitled 'Implementing Environmental Commitment by the Insurance Industry'. The resulting statement of commitment by 60 signatories reflected the industry's earlier assertions that environmental deterioration leads to more financial risk for insurance companies.[31] At the same time, the UN University sponsored the World Congress on Zero Emissions, financed by Dupont, Sodra Cell and Canon, Inc., which concluded that the development of non-polluting technology combined with entrepreneurship and good business practice will lead to the goal of zero emissions.

In 1997 and 1998 the WBCSD continued its involvement with the UN by participating in the *ad hoc* inter-session working group of the UN Commission on Sustainable Development in February and in the Earth Council's 'Rio+5' Forum. They also sponsored a joint seminar on business and industry with ICC. Working through the Forum's plenary session, parallel sessions and workshops, the primary policy emphasis of the World Business Council was on joint implementation and emission trading as mechanisms for solving cross-border environmental problems. Active involvement in these preparatory meetings for the General Assembly's Special Session on Rio+5 in 1997 is precisely how NGOs in general and the WBCSD in particular, influence the environmental agenda of the United Nations. The strategy was similar to the actions of the Council during the third Conference of the Parties in Kyoto in December where the main priority for WBCSD was on the removal of counter productive subsidies and of barriers to technology transfer, on government support in the form of business incentives for joint implementation projects, and providing greenhouse gas emission credits at the national and international levels.[32]

By June 1997 the WBCSD had organised a large corporate delegation to attend the special session of the UN General Assembly which met for the Earth Summit to review progress in implementing Agenda 21 of the 1992 Rio Conference. WBCSD delegates participated in a special roundtable meeting of government officials and senior UN representatives and also in the plenary session. Indeed, the appearance of Mr Eugenio Clariond Reyes (CEO of Groupo IMSA, Spain) and Mr David Kerr (CEO of Noranda and Vice-Chairman of WBCSD) represents the first time that business leaders have addressed a plenary session of the United Nations. UN Secretary-General, Kofi Annan, has given strong support to the notion that the UN should work closely with the private sector and that policy is now becoming more evident in several UN organs and specialised agencies. The focus of the 1997 General Assembly was on economic incentives for business to protect the environment and that idea is reflected in the WBCSD's support for Agenda 21 in exchange for conditions which remain important to corporations, e.g. international standards of equality, realistic targets, stable trade relations, open markets.

In addition to the activities of ICC and the WBCSD directly targeted at the development of regulatory policy through UN conventions, corporations have also adopted the strategy of supporting partnerships with UN subsidiary bodies and NGOs in developing countries. For example, Shell and the Global Environment Facility (GEF) have agreed to provide funds and technology transfer to a Biomass Integrated Gasification/Gas Turbine commercial demonstration project in Brazil. Of significance as a model of partnership, the agreement reflects corporate interest in market development linked with assistance to developing countries through the mechanism of the UN. The GEF, a partnership with the World Bank, UNEP and the UNDP, provides funds for projects in support of renewable energy technologies which appear close to commercialisation and which are suited to widespread dissemination.[33]

Collaboration between the Prince of Wales Business Leaders Forum, the World Bank and the UNDP in 1996 to produce a report on several such partnerships is illustrative of how important the new linkages have become. Japan's Nature Conservation Fund project aims to build a network of local NGOs, the World Bank and the World Conservation Union (ICUN) to create jobs in developing countries. This initiative is indicative of the trend to link corporate interests with UN sustainable development policies. Increased linkage also includes a role for environmental non-governmental organisations, especially those that are active at the local level in developing countries.

CORPORATIONS AND NGOs

Not-for-profit non-governmental organisations which focus on environmental problems have proliferated to an extraordinary degree since the 1972 Stockholm Conference on the Human Environment and a vast literature on their activities now exists.[34] As noted earlier, this literature tends to exclude industrial NGOs such as the International Chamber of Commerce and the World Business Council for Sustainable Development. Because the WBCSD was formed very recently and, in part, because there continues to be strong resistance to the notion that corporations can or will accept responsibility for protection of the environment, a paucity of attention has been paid to corporate–NGO collaboration. Some environmental NGOs distrust the activities of business in general and of multinational corporations in particular; others have recognised the value of co-operating to achieve common objectives. Those NGOs which view the resolution of environmental problems through the prism of radical social transformation and which adhere to anti-capitalist philosophies will continue to analyse corporate involvement in environmental issues, both national and internationally, from the perspective of deep suspicion and distrust. Their belief that the long-term strategy of corporations is to dilute the effects of regulatory policy in order to continue business as usual remains a fundamental obstacle to their participation in any form of partnership.

Trends in corporate–NGO partnerships

Two trends, however, have contributed to wider recognition by NGOs that corporations can contribute positively to sustainable development, particularly if NGOs maintain their watchdog function.

The first of these was the emergence of companies such as Out of This World supermarkets, The Body Shop in the UK, Green Power, Patagonia, Ben and Jerry's in the US, and Ecover in the Netherlands which adopted environmentally sound practices from the beginning. The second trend relates to the changes which have occurred in corporate attitudes toward environmental issues.[35] A debate has emerged in the literature of business studies as well as in international relations concerning the reasons for this change in policy. It is a debate which centres on corporate motivation. Are corporations trying to avoid negative publicity engendered by irresponsible practices, by the threat of fines and lawsuits, by the promise of new markets for clean technologies, by enforceable government regulation, by a genuine commitment to protecting the environment, or by a combination of several of these factors? The nature of the debate was clearly illustrated in a Cambridge University forum on exchange between NGOs and multinational corporations in which the central question of whether or not MNCs can serve as agents of sustainable development dominated the discussion.

Clearly the answer to that question lies in the perspective of individual corporations and NGOs, but what emerges from an analysis of the debate is the increasing trend of less developed countries toward market-orientated policies which, in turn, lead to greater reliance on the private sector.[36] Their expertise in project implementation, access to the latest technology, their ability to finance resources as well as their global distribution through local branches and their increasing direct investments in developing countries, makes MNCs a natural ally of governments and non-governmental organisations with limited resources. As has often been observed, corporate support is invaluable for environmental monitoring and for implementing community development projects as well as insuring compliance of their subsidiaries with the local, national and international regulation.

Whilst it is important to maintain a distinction between those NGOs who remain suspicious of business motives and those who view business as partners, it is nevertheless accurate to assert that there is growing co-operation between business and some non-governmental organisations. For example, the International Union for Conservation and Nature (IUCN), one of the oldest NGOs and one of the earliest to be given consultative status with ECOSOC, has recently issued a joint report with the World Business Council for Sustainable Development. The report, *Business and Biodiversity: A Guide for the Private Sector*, discusses a wide range of issues likely to be encompassed in international conventions and which will affect pharmaceutical, agricultural, biotechnological, mining, forestry and the petroleum industries in the future.[37]

The greening of Shell: working in partnership

Shell UK, after a remarkable reversal in environmental policy in the aftermath of the Brent Spar case and widespread criticism of its policies in Nigeria, is now working with the World Wide Fund for Nature to explore the issues raised by the disposal of the Brent Spar. It has also financed a series of reports produced in collaboration with WWFN on a range of economic, social and environmental issues which affect the development of forestry plantations which have been used to develop guidelines for best environmental practice in plantation forestry.[38] Shell is also monitoring the development of forest certification by the Forest Stewardship Council to assist the movement toward global environmental management standards within the ISO 14000 series. It is working with environmental groups in Peru to implement an environmentally sensitive exploration site at Camisea in Peru which may well become a model for environmental sensitivity and transparency, particularly since Shell's performance will be monitored by a Peruvian network of 35 NGOs called Red Ambiental Peruana.[39]

As Table 4.3 illustrates, the collaboration between corporations and NGOs is extensive, and there is an important trend in collaboration between some NGOs and some corporations. It is a trend which is clearly seen by both the specialised agencies of the UN and by developing countries as a practical solution to the potential conflict between the desire for environmental improvement and the lack of resources required to achieve that objective. It also demonstrates that with a balanced approach to interests and needs, collaboration which serves shared goals can and does occur.

Table 4.3 Environmental partnerships between MNCs and NGOs

MNC	*NGO*	*Project*
Glaxo Wellcome	Pro-Natura International and Fauna and Flora International with University of Istanbul	Tropical forest conservation and plant identification in Turkey.
B & Q	Forest Stewardship	Forest management.
NEC	Wild Bird Society of Japan	Monitor annual wild crane migration and wetland monitoring programmes in China, Korea, India, Mongolia and Russia.
Dow Chemical	Natural Resources Council (USA)	To achieve Dow's waste and emission reductions goals through the use of a third-party pollution prevention auditor.
Unilever	World Wide Fund for Nature	Support sustainable fishing through establishing a Marine Stewardship Council.
Dupont	University of Oviedo and several Spanish NGOs	Restore a wetland site, plant trees, support local environmental education, develop a ground water protection model and a zero emissions plan.
Procter & Gamble	National Audubon Society (US)	Managing wildlife habitat in private forests, provide training and education programmes for loggers and landowners.
Johnson & Johnson	World Wide Fund for Nature	Assessment of the trade in rhino and tiger body parts, generating data to support the Convention on International Trade in Endangered Species.
BHP Petroleum	Nature Conservancy (US)	Sale of Diamond Y Spring preserve in Texas to NC in order to protect wildlife, flora and fauna.

CORPORATIONS AND REGIONAL ORGANISATIONS

In addition to collaborating with international and non- governmental organisations, corporations have also focused on regional organisations which adopt and implement environmental policies. Substantial evidence of the partnership approach may be found in the European Union, the Association of Southeast Asian Nations (ASEAN) and the Organisation for Economic Co-operation and Development (OECD).

ASEAN's Southeastern Technical Advisory Committee inaugurated a water resource management system, which will involve governments, non-governmental organisations and the private sector. An important part of the regional environmental strategy is to transfer 'best practice' from one member to another, thus achieving savings in research and development, as well as reducing implementation costs. Specifically, assistance is being provided by the private sector in working out the details of Eco-transfers to subsidiaries or through joint ventures, in promoting dialogue and sharing information and in a commitment to help its members meet ISO 14000 series of standards.

The recently negotiated arrangements in North America provide a particularly good example of environmental regulations agreed between two developed countries and one developing country and implemented through a regional organisation which is responsible for compliance and in which corporations have been involved in shaping the agenda.

Monitoring the environment in the North American Free Trade Agreement

The North American Free Trade Agreement, in force on 1 January 1994, commits the United States, Mexico and Canada to a free-trade area and had its origins in two previous agreements: the 1987 US – Mexico Framework Agreement and the 1989 Canada–US Free Trade Agreements. In addition to NAFTA's core elements (dispute settlement; rules of origin; provisions for the automotive and textiles industries; rules on investment, financial services, government procurement), the three countries negotiated 'side-agreements' on the environment and labour in 1993. The largest North American automobile manufacturers (General Motors, Ford and Chrysler) were intensely involved in the NAFTA negotiations, particularly with respect to gaining access to the Mexican automotive market and to preventing Japan from using the NAFTA provisions to gain preferential access to the US market.[40] Likewise, PEMEX, the Mexican state-owned fuel company and its electric company (CFE) became involved over the provisions for non-discriminatory government procurement markets.

NAFTA's side agreement on the environment created a new permanent Commission on Environmental Co-operation to monitor compliance with the NAFTA provisions on the environment, which include sanitary and phyto-sanitary measures, harmonising standards to the highest common denominator and environmental criteria for investment. NAFTA's dispute settlement mechanism not only allows the Commission to investigate disputes and to increase transparency, but it also has the right to levy sanctions and it places the burden of proof on the party challenging the environmental standard. These efforts to contribute to sustainable development are supplemented by non-NAFTA agreements such as the 1992 US–Mexico Border Plan which addressed the border pollution problem through joint initiatives and stricter enforcement of industries located on the border.

In July 1997, the Commission on Environmental Co-operation released the first integrated study of Canadian, Mexican and US toxic chemicals released into the environment. The report, for the first time, includes transfers of products to recycling, reuse and recovery of energy as well as mandatory release and transfer data, and indicates that while the US leads in toxic chemical emissions, Canada continues to be the worst offender of discharges into surface water. Mexico, in 1992, implemented legislation creating the Federal Environmental Enforcement Agency which is now in the process of evaluating industrial compliance, and the first eighty firms have received clean industrial certificates. Nearly half of those companies receiving certificates were American or other foreign firms. In a co-ordinated action, the US Environmental Protection Agency has encouraged US border companies to participate in Mexico's certification programme, and its environmental audits will serve as the basis for fiscal incentives to corporations that undertake environmental re-engineering. Mexican, American and Canadian corporations were actively involved in the negotiation of these environmental agreements and they are now attempting to comply with the environmental regulations which have ensued. In those cases where resistance is mounted, the Commission can threaten and implement sanctions. NAFTA and the side-agreement on the environment provide yet another example of the complex linkages between governments and corporations that emerge in the negotiation of international and regional environmental agreements. It also demonstrates the extent to which co-operation between governments and corporations is essential for implementation of cross-border environmental agreements.

NAFTA's free-trade provisions also provide ample opportunity for Mexican enterprises to expand into the American market, especially with respect to 'green products'. As the American demand for organic food nears a rate of 20 per cent annual growth, Mexican food producers are now selling US$500 million worth of organic food products annually.[41] Mexico's organic food sector not only provides a positive example of a developing-country enterprise taking advantage of environmental market opportunities, but it also illustrates the opportunity and incentives which can be created by regional trade agreements and their subsequent institutionalisation.

THE GLOBAL TRIAD

This chapter has focused on the role of multinational corporations in international organisations and the challenge to traditional concepts of diplomacy, representation and sovereignty that the corporate role poses. In the dense web of interactions formed by governments, non-governmental organisations and corporations trying to shape environmental policy, it is not possible for any single group, including very large multinational corporations, to dominate the international environmental agenda. Corporations, acting through non-governmental organisations such as the International Chamber of Commerce and the World Business Council for Sustainable Development, have been restrained by the activities of governments and NGOs and they have also been able to provide significant leadership, innovation and resources in support of sustainable development. Intense interaction among the three groups, each pursuing resources and advantage, requires compromise and co-operation; in this sense, they form an important global triad.

5 Greenpeace: Case Study of an NGO Multinational Enterprise

Reality Field 5
Large, transnational non-governmental organisations are increasingly organised and operated like multinational business enterprises and have taken on some of the organisational characteristics of corporations.

INTRODUCTION[1]

The global triad composed of governments, corporations and non-governmental organisations (NGOs), as discussed in the previous chapter, relies for much of its strength and innovation on NGOs such as Greenpeace, World Wide Fund, Sierra Club and Friends of the Earth which have become international non-profit businesses. Indeed they have become million-dollar, sophisticated, multinational operations which engage in research, political lobbying, effective media communication and activist policies on behalf of the environment.[2] Two of the newer organisations, Greenpeace and Friends of the Earth, founded as 'lunatic' fringe groups by environmental activists in the 1970s, had by the end of the 1990s become major actors in their own right, both as non-governmental organisations granted observer status by the Economic and Social Council of the United Nations and as international enterprises with global reach. So successful were they at their debut in the development of international environmental policy at the UN Rio de Janeiro Environment and Development Summit in 1992, that they led a group of leading environmental NGOs in the conference negotiations.[3] As noted in the previous chapter, the international NGOs have played a significant role in shaping the emerging global environmental agenda and none more so than the most widely recognised NGO, Greenpeace.

Although it is customary to treat Greenpeace as a non-profit NGO, it may also be compared to pressure groups and, for the purpose of this analysis, to a small, but rapidly growing, international enterprise.

Despite its non-profit state, it shares the problems of creating new products and developing new success strategies with other growing businesses. It also reflects the importance of technological leadership in increasing its special competitive qualities. Beginning its organisational existence as an activist pressure group that campaigned for changes in the law to protect the environment, Greenpeace moved during the 1990s radically towards an approach which may be compared with other transnational enterprises. By 1995, for example, it had put in place telecommunications capabilities able to co-ordinate a complex international media and direct action campaign around the world. It introduced its first telecomputer system in the late 1980s, long before many similarly sized multinational businesses had begun to consider such technologies.

Greenpeace has been managed by a number of individuals who were successful business entrepreneurs prior to joining Greenpeace. Its aims are different from those of ordinary businesses in that it does not generate profit for shareholders. But many of the methods the organisation uses may be compared with other multinationals.

In 1995, Greenpeace had about £47 million of accumulated funds,[4] an annual expenditure of £90 million plus staff salaries[5] (1995), and 1,300 employees world-wide in 43 offices.[6] Today it operates globally, with offices in 32 countries and 2.9 million members in 158 countries. It writes business plans and has expanded its activities to newly industrialising countries with major environmental problems. Its annual income of approximately $125.6 million is generated by annual subscriptions, donations, fund-raising events and the sale of merchandise. Like many other enterprises, it has commercialised its brand name and in the right circumstances sells Greenpeace T-shirts, socks, CDs and banners in special shops in major cities. From time to time, it has offered the services of its professional employees as technical consultants to industry and to government, always with the proviso that the results of the consultancy can be published.

Growth to this size has meant that Greenpeace has had to hire management consultancies on a number of occasions to review its structure and strategy. Its marketing strategy has now been developed to take account of the fact that income drops in periods of economic recession – as in the case of the early 1990s. By the mid-1990s, Greenpeace's strategic analysts were recommending the replacement of revenue in boom times with more systematic and controllable investments and income sources. Greenpeace, like many other businesses, grew too fast in the late 1980s and then had to downsize and outsource, producing its range of goods through franchised contractors. This strategy has resulted in a good balance between central control in its Amsterdam headquarters, and local devolution. Thanks to the international telecommunications systems, complex decisions can be discussed, relayed and implemented rapidly. Therefore, it is able to implement world-wide policies, equivalent to standard 'products'. For example, its anti-whaling ban is applied in its Japanese and Norway offices, in spite of national opposition.

Due to its strategic orientation, it has had some notable successes. In 1984, it identified the export of toxic wastes from advanced to developing countries as a major issue. Through persistent pressure, publicity and behind-the-scenes negotiation, by 1994 it had successfully catalysed an international ban on this trade. With the advent of the World Trade Organisation in 1995, this ban has been permanently instituted. Thus, a major potential for gross environmental inequity – the rich imposing environmental hazards on the poor – has been made illicit with world-wide support from national governments. As with many multinationals Greenpeace perceives the rich countries as markets where its messages and consequent income streams have reached maximum penetration. In response to having reached this plateau, Greenpeace made the decision to expand the business to countries where the environmental message is still faint and marginalised. Besides China and India, the nations of the Russian commonwealth are obvious targets.

All the major polluting economies have growing numbers of educated professionals and a new generation of wealthy entrepreneurs. While many such individuals will have little interest in giving back some of what they have taken from their communities, the Greenpeace option provides a method by which relatively small amounts of money can generate substantial government and private sector action. Campaigns with proven effectiveness can be mounted. In an age of declining religiosity and church affiliation, some successful business people have seen organisations such as Greenpeace as secular equivalents to the churches of an earlier time. Rather than giving money to religious sects, Greenpeace and other environmental groups have provided a means by which successful entrepreneurs can share their wealth with a visionary cause which does not require any spiritual allegiance.

Critics charge that Greenpeace's environmentalism is simply a form of Western colonisation, comparable with McDonald's and Coca-Cola as a world brand with an 'imperial' agenda. Greenpeace's refusal to vary its campaign messages in responding to local community preferences, as in whaling and seal culling, is seen as confirmation of its essentially North Atlantic assumptions. Other critics have focused on Greenpeace's radicalism and portray it as an anti-industrial lobby group whose underlying agenda is to shut down industrial society.

This reality is a strategic challenge for the organisation's planning capabilities. So far, it has ridden the wave of environmentalist awareness with adroitness. A continued capacity to anticipate changing tides of environmental concern will be essential. This corporate revisioning must be progressive. It will condition whether Greenpeace can engineer itself to respond to new markets' and to younger generations' environmental preferences.

THE ORIGINS OF GREENPEACE

Greenpeace was founded by a small group of American and Canadian anti-nuclear activists living in Vancouver in 1971.[7] Irving Stowe, an American Quaker, and Jim Bohlen, a former US Navy diver, met at an anti-Vietnam protest in 1967, and in 1971 they were both living in Canada to avoid American conscription. Inspired by a Quaker group which in 1958 had sailed to the Bikini Atoll in protest against American atomic testing, Irving and Dorothy Stowe and Jim and Marie Bohlen along with Paul Cote and Bill Darnell, a Canadian social worker, formed a committee in 1971 called 'Don't Make a Wave' to organise a similar voyage into the US atomic test zone in Amchika, Alaska in an attempt to halt US underground atomic testing. This committee chartered a 24-foot halibut seiner (fishing boat) called the 'Phyllis Cormack', owned by Capt. John Cormack of Vancouver. Funds for the protest were raised by selling 'Greenpeace' campaign buttons and T-shirts; thus a powerful logo combining environmentalism and anti-war sentiment was born. Three journalists, Robert Hunter, an American anti-war protester with the *Vancouver Sun*, Ben Metcalf of Canadian Broadcasting Corporation and Bob Cummings of the *Georgia Strait* joined the crew, thus ensuring maximum publicity. However, the US government postponed the tests, winter arrived in Alaska and the voyage had to be abandoned. In the meantime, Irving Stowe managed to raise funds for a second boat, 'The Edgewater Fortune', to return to the testing site. Engine trouble intervened and the protesters were 1,100 km from the test zone when the bomb exploded. A year later, the US, for other reasons, decided to halt underground testing. Not discouraged by failure, the committee agreed to wage a campaign against France which was the only Western power still conducting tests above ground in 1972. An advertisement was placed in the New Zealand newspapers for someone who owned a boat and was prepared to sail to the French testing site at the Mururoa atoll in the Pacific Ocean.

David McTaggart, a Canadian expatriate who had been a building contractor in the United States in the 1960s, was living in New Zealand and volunteered to sail his private yacht 'Vega' to Mururoa to try to prevent French atmospheric nuclear bomb testing. This brave effort was foiled by the French navy ramming his yacht and then successfully conducting the intended test. However, McTaggart was back in Mururoa the following year (1973) when the French boarded his boat and gave him a severe beating for which he subsequently received compensation from a French court. With the death of Irving Stowe, Ben Metcalf became chairman of the association and led the campaign against French nuclear testing. The French did not abandon atmospheric testing until 1975, but the dramatic confrontation between Greenpeace and French warships attracted unrivalled media attention to the issue of atmospheric testing and, in this sense, the campaign had succeeded.

Sinking of the Rainbow Warrior, Auckland harbour, New Zealand, 10 July 1985

As part of its anti-nuclear stance in 1985, Greenpeace once again sailed its flagship, this time, *Rainbow Warrior*, to Mururoa atoll in the South Pacific in an attempt to disrupt an underground French nuclear test. En route to the atoll, the *Rainbow Warrior* was docked at the Auckland harbour on 10 July 1985 where it was bombed and sunk, killing a Dutch photographer. Two French security agents were charged with murder and five others were the subject of extradition to stand charges in New Zealand. As the dispute between France and New Zealand escalated over whether the convicted French agents should be imprisoned in New Zealand or repatriated to France, the latter imposed unofficial trade sanctions against New Zealand. The French action led to mediation of the dispute by the UN Secretary General Perez de Cuellar in July 1986 which provided for a French apology to New Zealand, the payment of $8.16 million in compensation and release of the two agents to a French military base in the Pacific. Lloyd Cutler, the American lawyer for Greenpeace, claimed that it was the 'first time a sovereign government had agreed to binding arbitration with a private NGO over a dispute concerning actions taken by its military forces'.[8] This episode established Greenpeace as a formidable opponent of nuclear testing and proved a powerful forerunner of the North Sea confrontation in 1995.

Shell Oil, Greenpeace and the Brent Spar drilling rig platform, 1995–8

Shell, the Anglo-Dutch international oil giant, discovered how the relatively small Greenpeace could impacts its core retail business. By the mid-1990s, Shell Oil had taken a leading role in developing corporate environmental relationships. Sponsoring many environmental awards and committed to international organisations like the World Business Council on Sustainability and the Environment (WBCSE), the company carried out environmental assessments and environmental audits in many local installations.

One of many such assessments was made during the 1990s in the North Sea oil fields, regarding the best available option for disposal of one of the huge platforms which were used to drill for oil in the previous two decades. The recommended option was that the least damage to the environment would be produced by disposal of the platform in a deep ocean trench. Here, a team of scientists and engineers advised Shell, it would slowly break up and its toxic residues would gradually spread across the deep ocean floor. The other option, to break up and dispose of the rig close to or on shore, exacted much higher environmental risks and would be much more expensive.

Greenpeace, meanwhile, had reached a point where its corporate maturity and the wide acceptance of its 'products' internationally had brought about a state of corporate stasis. The organisation since acquiring huge membership had seen a decline in expansion. Many committed activists had resigned; some were bored and a few distrusted the new 'business' image for the non-profit organisation.

Greenpeace faced a conflict. Would it settle into mature middle age as a gentle pressure group, having been accepted as an important non-governmental organisation (NGO) at the World Environment Summit of Rio de Janeiro 1992? Or would it renew the militant strategy of its youthful phase and recruit a new generation of risk-taking adventurers in the environmental cause who are willing to risk their lives in non-violent confrontation in front of the video and film cameras?

Brent Spar oil rig was not the ideal issue. But it was there. During the summer of 1995, Shell had announced in a low-key press conference, that the Brent Spar's massive skeleton would be towed into the deep reaches of the North Sea, dynamited and scuttled. In June, two massive sea tugs began to move the vast structure through seasonably calm seas. As they moved, they found themselves being shadowed by *Rainbow Warrior*, one of the several seagoing vessels of that name which Greenpeace had so effectively used in its anti-whaling campaigns of the mid-1980s.

In earlier episodes, the tactic was simple but dangerous. The *Warrior* placed two rubber dinghies with high-power motors between the whaler and the whales whenever the vessels – usually Norwegian or Japanese – were closing for the kill. The whales, alerted by the commotion, could then dive deeply or swim away. This time, Greenpeace had decided that the *Warrior* required a helicopter and volunteers to be lowered onto the Brent Spar platform. By the 15th of June, four volunteers were perched uncomfortably on the remnants of the huge platform that had once contained a hotel and working quarters for 140 men.

What would happen now? There were clear indications in the weeks afterward that the UK government had gathered onshore a small force of commando troops to clear the Greenpeace environmental commandos. Having accepted that the company had chosen the best option, the government was willing to use its armed force to ensure that within the law the company could carry out its intentions.

However, by now the world press had been alerted. The method was the secret: Greenpeace had established on-line links with the major wire services, like Reuters, using video cameras and its own high-technology satellite-based broadcasting gear, recently purchased. This provided the capacity to broadcast high-quality television images of environmental confrontation from the remotest places on earth. Instead of the old-fashioned paper press release, Greenpeace provided broadcast-quality film footage generating instant press coverage to the major news services at no cost to themselves. Any editor had to consider whether a story with such potent reader/viewer interest and obtained at no cost and with high technical quality, should run on the evening news.

In the case of the Brent Spar footage, this technological versatility and astute management of the TV news services worked superbly for Greenpeace's campaign managers. Within hours of the rig arriving at its seabed destination but still afloat, virtually every European mainstream broadcast organisation was covering headline images of the confrontation.

Within European Union countries, a wide awareness of the pervasive power of the oil giants, or Seven Sisters, as the world's largest energy companies are known, is well established. In Germany Greenpeace's story triggered a consumer revolt against Shell retail products, particularly gasoline. Consumers switched purchases from Shell to other petrol stations to show their distaste for the Brent Spar decision. Within a week of the first press and TV stories, Shell retail sales in Germany and other Continental countries had fallen by 30 per cent – a matter of hundreds of millions of dollars, far more than the cost of even the most expensive form of eco-disposal of the oil rig.

Suddenly the costs of carrying out a Greenpeace-preferred option, no matter how high, began to be dwarfed by the loss of cash flow. In a matter of days, Shell's main board in London had directed a change to on-land deconstruction of the platform. In an ironic footnote, a number of larger Germany shipbuilding consortia would be bidding for this work. The chairman of Shell appeared on national television in every country, acknowledging his company's misunderstanding of public awareness and concern.

In a final ironic footnote and only a few months later, Greenpeace's scientists acknowledged that they had got their estimates of seabed pollution wholly wrong. But by then the oil rig was anchored in a Norwegian fjord awaiting further instructions as to its ultimate fate. The final decision by Shell for the disposal of Brent Spar came on 29 January 1998 when it was announced that the Spar would be disposed ashore in Norway and would be made into a new quay extension at Mekjarvik. The topsides will be scrapped for recycling. Given the presence of over 100 oil installations in the North Sea which will eventually require disposal, the Shell decision will become a model of disposal for others to follow.

Another important outcome of this case was that the relations between Greenpeace and the media, particularly in Britain, significantly worsened. The media charged that it had been manipulated by Greenpeace, had been given false data and that its reputation for accurate reporting had suffered. In turn, Greenpeace brought a complaint to the Independent Television Commission against Channel 4 for misrepresenting the nature of a BBC interview with Greenpeace UK Director, Peter Melchett, and later for its bias in a programme on BBC2 called *Scare Stories: Going Overboard* which was aired in 1997. From the perspective of Greenpeace, the Brent Spar case was a clear victory which changed the direction of policies on ocean dumping, forced Shell to retreat from its decision and which was subsequently undermined by Shell's success in persuading the media and the UK government *ex post facto* that Greenpeace had acted irresponsibly by providing video footage to the media and by its calculation error. From the BBC's perspective, the admission of error by Greenpeace, whilst 'courageous and honourable', had caused the public and journalists to be more sceptical of Greenpeace's work.[9] Acting in tandem with the BBC, the UK government claimed that the media had been co-opted by Greenpeace into misleading the public and the government. Greenpeace, however, received strong support from other environmental groups (Friends of the Earth, World Wildlife Fund UK, The Wildlife Trust) who also believed they had been deceived by the BBC regarding the true intent of its programming.

It is perhaps too early to attempt a balanced assessment of who manipulated whom in this affair, but for an organisation whose reputation depends on media coverage, any loss of credibility with mainstream journalists would be serious. That Greenpeace was concerned about this loss of credibility is made abundantly clear in the 1998 Greenpeace publication, *The Turning of the Spar*, which details every attempt made by Greenpeace to refute the charges against it. In the long term, however, Greenpeace will be remembered for having prevented a very large oil company from dumping pollutants in the sea, not for its dispute with the BBC and the UK government in the aftermath.

As the previous two cases illustrate, Greenpeace, for the past three decades, has conducted campaigns on both environmental and nuclear issues. Publicity from these campaigns helped expand Greenpeace into a multi-million-pound leader in environmental and antinuclear activism. Through a series of courageous direct action policing campaigns, it has scored major successes, including publicising French atmospheric nuclear bomb testing in the Pacific, challenging Japanese policy on drift-net fishing and smuggling of endangered species, changing Japanese and Norwegian whaling policies, raising awareness of the dangers of dumping radioactive waste in the oceans, of ozone-depleting gases, of greenhouse gases and the threat of de-forestation on a global scale. Table 5.1 summarises these campaigns.

Its best year to date was 1991, when it had 4.8 million members world-wide and accumulated cash reserves of over £50 million. Since then membership has slipped back to 2.9 million and its revenue

Table 5.1 Three decades of Greenpeace activism

	1970s	*1980s*	*1990s*
Environment issues	Whale hunts Seal hunts Over-fishing	Whaling Mineral exploitation Driftnets Clean energy Clean water Export of toxic waste to the Third World CFCs Dumping toxic waste Incineration of toxic waste at sea	Rainforest protection Clean energy Seismic surveys for oil exploration Whale & dolphin survey World heritage parks Green products Brent Spar Genetically-engineered food
Nuclear issues	Atomic testing	Radioactive waste at sea Underground nuclear tests Chernobyl Nuclear Free Seas Campaign	Nuclear testing Nuclear and toxic dumping

Sources: Greenpeace, *25 Years of Environmental Activism*; Greenpeace *Overview*, March 1997; Greenpeace *Campaign Report*, Summer 1998.

declined from $139,895 million in 1996 to $125,648 million in 1997.[10] In the Shell Oil case, a major multinational enterprise was presented in the media as substantially departing from its stated environmental ideals both in the North Sea and in Nigeria. Within a matter of days, the company was forced to restructure a key programme for disposal of a North Sea oil rig in order to regain market share in the European gasoline purchasing market. More importantly, consumers showed they were willing to alter their preferences to show their environmental concern.

This process of environmental visibility is as yet poorly understood, even by the most sophisticated of international companies. For the first time, the globe is united in holding responsible for environmental performance the most powerful economic institutions, the major corporations. Strategic awareness within these companies is only now beginning to embrace this process. This chapter seeks to understand Greenpeace's current position among such companies and the changes which have occurred in its basic strategy and mission as well as to ask some questions about how it can develop a viable growth strategy for the future.

THE ENTERPRISE: GREENPEACE

Structure

In its incipient stage, Greenpeace suffered from a lack of clear direction and unified leadership. Rivalries between David McTaggart in Europe and Ben Metcalf in Vancouver in the mid-1970s, the formation of several splinter groups including one led by Robert Hunter in 1972 to conduct the 'Save the Whales Campaign' and the death of Irving Stowe led to management fragmentation. From 1979 when McTaggert was appointed chairman of Greenpeace International until his retirement in 1991, a modicum of stability was established by uniting the offices in Europe, the Pacific and North America to form Greenpeace International.

Since 1989 Greenpeace International has been based in Amsterdam and is currently headed by Thilo Bode, its international executive director who is the third director in eight years. In 1998 Greenpeace International co-ordinates and is financially supported by national offices in 32 countries which contribute approximately one-quarter of their gross income to international campaigns and which send representatives to the annual general meeting. In addition to the national offices, there are four regional offices for Latin America (in Brazil), Central America (in Guatemala), the Mediterranean (in Malta) and Europe (in Brussels). Of the 32 national offices, 20 (or 63 per cent) are located in advanced industrial countries. The other offices are located in Latin America (Argentina, Brazil, Chile, Mexico), Eastern Europe (Czech Republic, Slovakia), the former USSR (Russia, Ukraine), Tunisia and China.[11] An international board, elected by the national offices, meets annually and sets the campaigning priorities for the forthcoming year.

As an article in *The Economist* observed a few years ago, Greenpeace

> has managed to pull off two balancing acts better than many multinationals. The first is between devolution and centralisation. Most decisions are taken locally. National offices, rather than Greenpeace International's headquarters in Amsterdam, run campaigns on national pollution issues, and are responsible for building contacts with national politicians and journalists. Yet when it needs to, Greenpeace can act like a centralised organisation.... Although bosses of national offices are given relative freedom from

> Amsterdam, they are never allowed to change Greenpeace's worldwide policies to suit local tastes. Greenpeace Norway, for example, is obliged to oppose whaling in spite of the howls of local fishermen. The Greenpeace brand is identical the world over.[12]

It is possible that this balanced structure is an outgrowth of Greenpeace's evolution from a small association based in one country to an extraordinary proliferation of semi-autonomous national offices, formed by strong and committed individuals, in a number of countries. Initially most of the so-called 'national' offices were simply a few people in an office but with access to the expertise, logos and reputation of Greenpeace International, much like a typical franchising operation. Initially they were supported financially by Greenpeace in North America, but increasingly were expected not only to become self-supporting, but to contribute to Greenpeace International as well. In this sense there has been a reversal in structure over the thirty years of existence since Greenpeace in Vancouver hired McTaggert's yacht in New Zealand.

Aims

Greenpeace is in the business of environmental campaigning. In its homepage on the Internet it describes itself as 'an independent, campaigning organisation which uses non-violent, creative confrontation to expose global environmental problems, and to force the solutions which are essential to a green and peaceful future'. Greenpeace's goal is to ensure the ability of the earth to nurture life in all its diversity. Therefore Greenpeace seeks to:

- Protect diversity in all its forms.
- Prevent pollution and abuse of the earth's ocean, land, air and fresh water.
- End all nuclear threats.
- Promote peace, global disarmament and non-violence.

More recently, Greenpeace has emphasised the Quaker principle of 'bearing witness' which was evoked by Irving Stowe and others in the earliest days of Greenpeace. Essentially it means being present where damage is being done and stopping it by simply being there. Rather than Ghandian non-violence, it serves as the philosophical foundation for Greenpeace's non-violent approach to action.

Greenpeace's aims, while appearing to be consistent since its inception, have actually received different emphases over time, depending on current issues of the day as well as the location of specific national offices. For example, Greenpeace in Canada is more forcefully involved in opposition to the logging industry than Greenpeace UK, and Greenpeace International was more involved in nuclear-testing issues in the early 1970s when the focus was on the United States and France than it is today where the testing programmes in India and Pakistan have not been the target of Greenpeace activities. Indeed the only direct action taken against nuclear proliferation in the sub-continent was flying a hot-air balloon with a banner attached over the Taj Mahal whilst the G8 Foreign Ministers were meeting in London. Greenpeace has been subjected to criticism that it monitors environmental harm more actively in developed countries than it does in developing countries. Certainly the emphasis on developed, Western countries is apparent in the list of national offices where only 8 out of 36 offices are located in developing countries and there are none in Africa.[13]

Currently the weight given to environmental damage from non-nuclear sources is much greater than that given to nuclear damage; indeed, since the 1970s, Greenpeace has not conducted a major campaign on nuclear issues and has been increasingly swept up in addressing issues of mineral exploitation, forest cutting, toxic waste dumping, fishing practices, the ozone problem, renewable energy and, more recently, genetically engineered food crops. The Greenpeace World Wide financial statement for 1997 indicates an expenditure on the Nuclear and Disarmament Campaign of $8.4 million which represents only 18 per cent of the total campaign budget. It appears to be the case that Greenpeace campaigns are being driven, in part, by NGO competition and by the development of UN environmental treaty-making.

It is also the case that Greenpeace action campaigns, in practice, have moved from targeting governments to targeting Western-based multinational corporations, perhaps because governments are less susceptible to the pressure of boycotts and unfavourable publicity than businesses; perhaps because the issues determine the target – for instance, only governments control nuclear facilities – and perhaps because Greenpeace is reluctant to attack developing country governments, despite their increasing responsibility for pollution, both nuclear and non-nuclear.

Legal status

Greenpeace International is fundamentally a not-for-profit organisation which has been recognised by the United Nations as a non-governmental organisation under the procedures previously described in Chapter 4. In the UK, the organisation is further divided into two branches. The first, Greenpeace Ltd, is responsible for campaigning, direct action, political lobbying and scientific research. In 1998 this branch was directed by Lord (Peter) Melchett. The other branch, Greenpeace Environmental Trust, is a registered charity which funds scientific research and carries out educational projects on environmental issues.

In a letter to the *Financial Times* on 21st September 1995 Lord Melchett, the Executive Director of Greenpeace Ltd, UK, denied that Greenpeace is a charity:

> In fact 90 per cent of our income (and our activities) are not charitable under UK law. In addition Greenpeace never takes money from governments or industry anywhere in the world (we have offices in more than 30 countries). Greenpeace campaigns in the UK are undertaken by Greenpeace Ltd, which is not a charity. The Greenpeace Environmental trust, which is a charity, funds charitable projects, mainly on research issues.[14]

The legal status of the organisation has clearly evolved with its institutional development from a private association of a few individuals to a multinational enterprise which operates across national boundaries, collects large sums of money from its subscribers and develops its own product line. As such it represents the legal complexity of international environmental organisations, if not quite the scale of large, profit-making enterprises such as Shell. Greenpeace, through its high-profile challenges to government in the European court and to companies through national courts, is now seen as a major player by both governments and corporations. For example, British Nuclear Fuels was challenged by Greenpeace in their operations of the Thorpe nuclear reprocessing plant in Cumbria requiring several millions of pounds in legal fees. There is always, even for a trust, the threat of seizure of assets. Unfavourable publicity can dry up funding. For this reason Greenpeace has decentralised and hybridised its company structure to minimise these threats and its liability.[15] Imprisonment of key executives and field team leaders could seriously hamper operational effectiveness.

Paul McGhee, who had been the London-based co-ordinator of international action, 'suggested the organisation was trying to combine publicity "hits" and more talks with big business and spend less time on long-term direct action'.[16] Whether McGhee's criticism is accurate or whether Greenpeace is simply combining two different tactics for the same strategy, what is clear is that Greenpeace is not a conventional business. Although income is generated from the sale of products as well as through individual subscription, part of the organisation functions as a charity.

Assets and finance of the enterprise

Greenpeace as an organisation had a long-standing theory that its enemies – which include from time to time armed forces of sovereign states, major multinational corporations and large government agencies – are so powerful that its funds must be dispersed and hidden in ways that cannot ever be accessed by its opponents, singularly or in combination. Consequently, funds are held in a network of trusts.[17] Limited companies and charities in each of its country operations clothe the operation in some anonymity. In the past, most of these were exempt from full disclosure and, as a result, some concern over inappropriate use of funds has occurred. For example in the UK, Greenpeace Communications Ltd operated on a grant from other Greenpeace organisations and was not individually identified in the accounts. The Greenpeace Environmental Trust funnels subscription funds into research. In a number of locations, including London, Greenpeace has built its own offices to demonstrate 'green' building technologies; but also perhaps to give it some real-estate-based security in the face of declining growth rate in subscription funds. In 1995 Greenpeace was accused of venturing into property deals using subscription funds (£3 million), of spending excessive amounts on the 1994 annual general meeting ($150,000) and of providing the Board with an expenses budget of $400,000.[18] It was also noted that no Greenpeace company is allowed to carry a long-term debt.

As of 1996, however, Greenpeace has been more transparent with respect to financial disclosure, but the details of its main source of income, presented in the financial statement as 'Grants and Donations', is still hidden, as is the 'Other Grants and Donations' category for the national offices. Its auditors, KMPG Accountants in Amsterdam, were confident enough of this transparency, however, to issue the following statement with respect to the pro forma financial statements of Greenpeace International: 'In our opinion, the pro forma summary financial statements set out on this page (Greenpeace web site page) are consistent, in all material respects, with the financial statements from which they were derived.'[19]

As of 1998, Greenpeace has three separate financial auditing categories which are published on its Internet website: Greenpeace International, Greenpeace World Wide and Greenpeace National Offices. The first, Greenpeace World Wide, is unaudited and indicates a world-wide income as of the year ended 31 December 1997 of $125.6 million and total expenditures of $128.0 million, leaving a deficit of $2.3 million, compared to a deficit of $5.4 million in 1996. Its Fund balance of $103.5 million was down from the previous year of $121.4 million, although this Fund represents a substantial reserve.[20] In the same year, Greenpeace International was shown to have a total income of $31.9 million and total expenditures of $30.3 million indicating a surplus of $1.6 million. Its closing Fund balance was shown to be $17.3 million.

By contrast, the 1997 total income of the UK office was $9.7 million, total expenditures $9.2 million, showing a surplus of $543,000. The UK 'Charitable Foundation' income was $1.7 million, of which $1.6 million was given to Greenpeace International. Greenpeace International gave a grant to the UK office in the amount of $151,000 (less than 1 per cent of the UK total income) and the UK office and its charitable foundation combined gave a total of $2.9 million to Greenpeace International. (26 per cent of their combined total incomc). Table 5.2 and 5.3 summarise the three statements, Greenpeace International, Greenpeace World Wide and Greenpeace UK, for comparative purposes. As these figures indicate, Greenpeace is still heavily dependent on annual renewals and recruitment to maintain membership levels and its income. These requirements constrain policy, dictate activities and make any changes in policies risky. What the figures also indicate is that Greenpeace is a wealthy organisation with substantial assets and cash reserve.

In addition to its financial resources, Greenpeace has physical assets in the form of a fleet of six ships (*Arctic Sunrise*, *Beluga*, *Greenpeace*, *Moby Dick*, *Rainbow Warrior*, *Sirius*), ranging from 23 to 58 metres in length, a hot-air balloon and an action bus equipped for exhibitions and campaigning, as well as real-estate assets in the form of national offices, e.g. the new Greenpeace building in London.

Table 5.2 Income and expenditures of Greenpeace UK, Greenpeace International and Greenpeace World Wide, 1997 (thousand US$)

	GP UK	*GP UK, Charitable Trust*	*GP International*	*GP World Wide*
Income:				
Grants from GP International	151	0	Not applicable	Not available
Grants from national offices	Not applicable	Not applicable	31,640	Not identified
Other Grants & donations	9,499	1,712	177	120,816
Interest	21	25	99	3,206
Merchandising & licensing	99	(3)	Not applicable	1,626
Total income	*9,770*	*1,734*	*31,916*	*125,648*
Expenditure:				
Grants to GPI	1,314	1,643	Not applicable	Not available
Grants to national offices			5,484	None
Campaigns*	3,079	40	10,339	46,137
Campaign support**	1,336	30	7,226	36,912
Fund-raising	2,713	7	1,315	23,857
Administration	785	48	5,952	21,118
Total expenditures	*9,227*	*1,768*	*30,315*	*128,024*
1997 position	In surplus	In deficit	In surplus	In deficit

* Includes biodiversity, toxins, climate, nuclear and disarmament campaigns.
** Includes media and communications, marine operations and action support, public information and outreach; except GPI which shows no expenditure for public information and outreach.

Sources: Greenpeace National Offices summary income and expenditure statements; Greenpeace International pro forma summary financial statements; Greenpeace World Wide pro forma summary financial statements, 1997. (www. greenpeace.org/report97/)

Greenpeace is currently the world leader among environmental campaigning organisations, with unrivalled experience in direct action. This experience curve, however, has been expensive for Greenpeace. In 1984 while it had approximately £30 million of assets in its ships

Table 5.3 Fund balances for Greenpeace UK, Greenpeace International and Greenpeace World Wide, 1997 (thousand $US)

	GP UK	*GP UK Trust*	*GP International*	*GP World Wide*
Opening fund balance	(693)	600	17,453	89,245
Closing fund balance	(130)	540	17,315	76,410
Total liabilities and fund balance	–	–	27,047	103,554

Sources: Greenpeace National Offices summary income and expenditure statements; Greenpeace International pro forma summary financial statements; Greenpeace World Wide pro forma summary financial statements, 1997. (www.greenpeace.org/report97/)

and related communications systems, a major campaign costs Greenpeace £1.5 million, excluding the use of existing assets. Its running costs are approximately £10 million per annum, £30 million for capital ships, special purpose fitting out, wharfage and another £10 million for offices, fitting out, IT and communications equipment. These cost estimates were calculated in 1995 and are almost certainly higher now.

In the mid-1990s, Greenpeace was accused of financial scandals over the misuse of funds and a spate of articles in the UK press highlighted its lack of fiscal transparency.[21] In 1997–8, the new open public image created by the publication of financial pro forma statements on the Internet was presumably the result of media criticism of its fiscal practice and procedures. And, in spite of the new transparency, the national accounts and GP World Wide accounts remain unaudited.

Greenpeace products and marketing

Some products, such as T-shirts, books, reports, campaign buttons, can be sold directly. Other 'products' such as political lobbying, media and public information are indirect and present a greater challenge to assessing overall marketing costs. For the purposes of this analysis, Greenpeace's products can be listed as:

- Direct Action
- Legal Action

- Political Lobbying
- Branding
- Greenpeace products from the marketing outlets such as T-shirts, etc.

Because of the particular accounting categories adopted by Greenpeace in 1996 and 1997, the total marketing budget for these products must be estimated. Using a definition of product which includes expenditures on media and communications, public information and outreach as well as marine operations and action support, Greenpeace World Wide has a substantial marketing budget which represents 29 per cent of its total expenditures. Greenpeace not only exercises a high degree of control over media exposure. It also seeks to control communications about itself, knowing the power of mass communications to alter public perceptions of its 'brand'. Retaining a high media profile assures the success of the campaigns on which GPWW spends 36 per cent of its budget.

For Greenpeace International in 1997, expenditures for media and communication alone were $1.2 million, down from $2.7 million in 1996. It is not clear from the new accounting format adopted by Greenpeace in 1996 how much of the administrative budget is spent on direct mail and advertising, but estimates made in 1995 suggested that these expenses were substantial. It is possible that these expenses are now part of the fund-raising budget, and for the purposes of clarity our analysis is limited to the figures presented in the 1997 pro forma summary financial statements. For Greenpeace World Wide, the expenditure on media and communications is $10.8 million, down from $16.4 million in 1996. If the expenditure for public information and outreach in 1997 ($15.7 million) is considered to be part of the marketing budget, the figure rises to $26.5 million. If Marine Operations and Action Support are also included in the marketing budget, the total is $36.9 million, as Table 5.4 illustrates.

Technology, telematics and information systems

Greenpeace is in the business of increasing public awareness of issues relating to the environmental health of the planet. Its innovative media applications in satellite-based communications within the field known as telematics, two-way radio telecomputing communications and control systems as well as tracking systems, has given it a

Table 5.4 Comparison of estimated marketing budgets for Greenpeace UK, Greenpeace International and Greenpeace World Wide, 1997 (thousands of $US)

	GP UK	*GP International*	*GP World Wide*
Media and communications	255	1,224	10,898
Public information and outreach	631	Not available	15,732
Marine operations and action support	450	6,001	10,282
Totals	1,336	7,225	36,912

Sources: Greenpeace National Offices, summary income and expenditure statements (unaudited), Greenpeace International and Greenpeace World Wide pro forma summary financial statements, 1997.

high-tech profile.[22] Accurate, detailed and immediate communications links are vital in order to provide as much global information as possible. Such operations are costly so any method of reducing costs is beneficial to the organisation. World-wide coverage allows increased communications efficiency and the ability for the organisation to highlight problems anywhere.

In addition to manually typed information, the use of environmental sensors and probes allows data to be automatically and continuously transmitted such as position, fuel consumption, status of cargo, etc. without the necessity of verbal communication. The use of placed probes allows continuous monitoring of radioactivity or pollution in a certain area. The use of sonar probes would allow monitoring of marine life or even marine mapping.

It also uses the new 'squisher box' for satellite communications, which allows for world-wide transmission of high-quality colour and sound pictures on the Internet. Its use of Internet allows for a wide variety of interaction with its various subscribers, supporters and critics. The websites allow better monitoring, although monitoring of remote environments remains labour intensive.

Greenpeace seagoing ships are well-resourced for communications and for research and testing. For example, *MV Greenpeace* is strengthened against ice and equipped with a helicopter pad. *Beluga* is fitted with a lab for pollution testing in rivers and coastal waters. The larger ships are all equipped with complete satellite communications facilities and tracking systems which allow them to follow the route of nuclear containers vessels. This system is appropriate for global co-ordination, good communications and rapid response, an advance in high technology which is perceived as important to the 'product' of direct action.

Management recognises that a large part of Greenpeace's success is due to its communications strength; thus making the maintenance of this strength a necessity. Greenpeace remains a market leader with high income in an under-developed marketplace which is expanding at a rapid rate, as the above-cited budget figures on campaign support indicate. As technology has moved on, Greenpeace seems to be keeping abreast of developments in telematics.

Human resources: how important is a decline in support?

In one sense Greenpeace's customers are everyone affected by its campaigning. Taking Greenpeace's successful campaigning against the French atmospheric nuclear testing in the early 1970s as an example, Greenpeace's customer base would include the whole world's population who have not been subjected to the resulting fall-out. However, in the business sense, Greenpeace's customers are those who subscribe to Greenpeace or provide financial support in other ways

According to Robert Rice, Greenpeace's membership peaked at 4.8 million in 1990.[23] In 1995 it dropped to 3.1 million and in 1998 it stood at 2.9 million, representing a decline of 40 per cent in eight years. In the UK office in 1995 there were 350,000 supporters, a drop from 400,000 at its peak, and in 1998 it had fallen to 215,000. Membership is directly linked to income, and by 1995, Greenpeace UK had suffered a 60 per cent fall in revenue. Evidence of declining membership suggests that Greenpeace may be in long-term decline. However, as Philip Kotler explains:

> management should not inevitably conclude that the brand is inevitably in the decline stage. If management withdraws funds from the brand it will create a self-fulfilling prophecy that will continue the brand's decline. Instead management should examine all the ways it could stimulate sales: modifying the customer mix, the brand's positioning, or the marketing mix. Only when management cannot identify a promising turnaround strategy might it conclude that the brand is in a terminal stage of its life cycle. Then it must decide what to do.[24]

Johnson and Scholes argue that the remedy for corporate decline is re-engineering the business.[25] One way Greenpeace has responded to this decline is to restructure its view of mass membership and direct action. In the mid-1990s, access to experienced activities was perceived as declining, as the non-activist membership, primarily engaged in fund raising, increased. For some long-term employees this mass marketing of the organisation has led to low morale. According to Peter Melchett, job cuts were made to finance the losses that were incurred from the reduction in fund-raising activity which, in turn, caused a loss of morale.

Because of lost support, Greenpeace sees a need to re-engage with ecological extremists. Local groups are to be encouraged to do less fund-raising and to take part in direct action. To this end training is being provided in non-violent protest techniques. Compassion fatigue, too, presents a threat, as the public develops an apathy towards environmental issues which remain unresolved. Overexposure is as constant a threat as underexposure. However, there is also evidence that the general public is increasingly aware of the need for long-term changes in ecological relationships and thus there may be greater potential for public action. The challenge for Greenpeace is to harness growing single-issue activists, e.g. road protesters, such as 'Swampy' in the UK, and criminal justice bill and animal rights demonstrators, whose energies and dedication are characteristic of the early Greenpeace activists. One possible threat to Greenpeace may stem from a long history of leadership rivalries, strong egos and disaffected employees, although the impact of these internal conflicts may have been mitigated by the strategy of national decentralisation adopted in recent years.

In the mid-1990s, there seems to have been a certain amount of disunity, with significant policy divisions at the executive and management levels, primarily on issues of secrecy vs. openness and the question of future strategy. Three top executives were ousted, taking with them large cheques and a confidentiality agreement. Since that time the leadership has been more stable.

Greenpeace and its competitors: the politics of environmentalism

The public is not obliged to support Greenpeace or any other environmental campaigning group. It could decide to spend money on any other product or service. Thus Greenpeace must constantly compete with all other opportunities for people to spend, save or invest their money. The more specific reality of competition among environmental NGOs for support in the form of subscribers and donations forces Greenpeace to adopt new and striking publicity campaigns so that they can remain in the public eye. Competitors include Friends of the Earth (FOE) and the Green Party. FOE is important in the UK but does not compete internationally. The Green Party is largely active in continental Europe and seeks environmental change through democratic means rather than direct action. In the UK a move towards proportional representation, currently being considered by the Labour government, could increase the influence of the Green Party. Moreover, the 1990s has seen an increase in single-issue political activities, e.g. road-building protests, animal-rights protests and opposition to genetically modified food by individual radicals. In France, the Netherlands and Germany, for instance, green parties are directly represented in national legislatures. Some mainstream parties in the UK and elsewhere in Europe are increasingly aware of the desirability to be seen as 'green'. Greenpeace needs to maintain a positive high-media profile to ensure that it is seen as separate from other environmental initiatives.

Greenpeace has independent political clout particularly on the basis of boycotts and campaigns which are specifically product-orientated. Moreover, Greenpeace has unrivalled experience in international environmental campaigning. This experience curve is expensive to maintain, but Greenpeace has substantial resources, especially excess cash, to fight off potential competitors. If necessary it could fund several more high-profile protests. Despite the operating budget deficit which Greenpeace World Wide shows for 1997, the Fund balance of $76.4 million and total assets of $103.5 million suggest that financial constraints are not a serious problem for Greenpeace. Furthermore, Greenpeace has independent political clout, particularly through boycotts and campaigns which are specifically product-oriented. Increasingly, companies and governments recognise that to be associated with green groups gives them legitimacy.

As Greenpeace does not depend on a few purchasers of its services and as it receives contributions and support from around the world, a more likely threat to its long-term growth would be public exhaustion and disinterest in environmental issues, media attention captured by radical eco-warriors who lodge themselves in trees and tunnels, charges that Greenpeace has become 'big business' and spends most of its time raising funds or by increased sensitivity to environmental pollution from both governments and corporations, thus robbing Greenpeace of its central mission.

Culture and vision: style and substance

High-profile campaigning by Greenpeace has always been coupled with risk through physical danger. There also continues to be risk through potential loss of assets – ships, such as the *Rainbow Warrior*, sunk by French military in the Pacific in the 1970s – and through false or misleading information such as that which occurred in the Brent Spar case. As the Greenpeace membership grows older and less radical voices dominate the policy debate, the culture of activism may shift to other, more radical groups.

In the past, the success of this culture was evident in the long-term strength of its popular appeal and it allowed the organisation to recruit youthful activists. Thus, Greenpeace was able to position itself as the Robin Hood of environmental campaigners. And therein lies the dilemma: if a protest is not dangerous and does not photograph well, it may not be enough for its 'customers'.

These methods are causing strategic problems at the moment. Campaigns are essential for publicity and are intrinsic to the image of the organisation, but increasing environmental awareness in the business field makes the need for such action less necessary. Governments, too, are becoming far better at anticipating and working with environmental protest, as are multinational corporations.

Hence there is substantial risk that the special 'niche' that Greenpeace has made its own may at least in advanced countries be evolving. However, there are many environmental challenges in developing countries where there is no shortage in the foreseeable future of dangerous confrontation and 'blackhat' environmental pirates. According to a former Director of Friends of the Earth, however, 'The secret of Greenpeace's success is that it has always looked like David against Goliath.'[26] According to Sam Warbrook, a BBC news editor, 'This particular David isn't armed with a slingshot so much as an AK47.' For example, he pointed out that Greenpeace is able to outspend television companies in shooting footage of protests which is then given to broadcasters free of charge. Given the size, global reach and wealth of Greenpeace in 1998, any attack on poor countries will appear to be the reverse of the David–Goliath image. Whether or not Greenpeace will be able to mount effective challenges against Third World and non-Western governments and businesses which pollute remains uncertain; its responses to Indian and Pakistani nuclear testing and to the appalling pollution levels in China and the former Soviet republics, as well as its current campaign budget allocations, do not bode well for this strategy.

Brand image and the power of instant global exposure

The strong brand image of Greenpeace, without doubt suffered some damage to credibility through its Brent Spar miscalculation. However, not to have admitted fault would have posed two greater threats: its own value of transparency would have been violated and it might have been exposed through inside information leaked by dissident employees.

Thilo Bode, Greenpeace international executive director, made this comment on the loss of the communications vessel *MV Greenpeace*:

> Our images are our strength, and it was very clever of the French to take that away... Donations will flow to the organisation that mounts the most high profile campaign, gets on the news most often and provokes the French commandos into the most violence.

Greenpeace does not operate on the basis of conventional exchange of goods and services for money. It carries out campaigns and hopes that these will bring in revenue as subscriptions and donations. It not only exercises a high degree of control over media exposure; it also seeks to control communications about itself, knowing the power of mass communications to alter public perceptions of its 'brand'. Retaining a high media profile means that it must at the same time not become overexposed.

As noted in the previous chapter, global broadcast media have become key 'enforcers' of good corporate practice. By presenting to world-wide audiences stories which indicate local deviation from advertised ideals, the press and television networks have the power to bring about a global levelling-up of corporate behaviour. Greenpeace's stated mission is 'to expose global environmental issues, and to force the solutions which are essential to a green and peaceful future'. But it is accused by some journalists of being a commercial concern, and one which, in effect, 'spends all its money in raising more money'.[27]

Strategic issues

Greenpeace enters its fourth decade with enormous achievements behind it. Yet similar to many maturing businesses it faces key strategic issues, some of which include the following:

1. What are new lines of work which will allow Greenpeace to regain and sustain its long-term growth pattern? It is our view that Greenpeace must adopt a new product or service to enter a new growth phase.
2. How can these new 'products' be made consistent with its established culture and vision of itself?
3. To what degree can Greenpeace become an ordinary 'multinational business network' without endangering the strengths which provided its early growth and success?
4. How can Greenpeace develop a truly global business for itself which can operate effectively in non-Western developing countries?
5. Given the global reach and wealth of Greenpeace combined with its new fiscal transparency, how can the organisation retain the 'David fighting Goliath' image which accounts for much of its success? Will it maintain the confidence and support of the national offices which exist as the primary fund raisers for GP International and which receive very little in return?

Answers to these questions are, of course, unknown, but it is certain that they will be determined as much by external opportunities as by internal decisions. Greenpeace will continue, like most successful corporations, to be driven by its market and by the wishes of its stakeholders.

Part II

The Meso-Micro Terrestrial Environment: Corporations, Nations and Cities

Part II addresses the complex, multi-layered nexus among corporations, nations and cities in the particular context of urban regeneration, corporation innovation, entrepreneurial approaches and sustainability. Chapter 6 surveys the themes of the corporate role in the alteration of metropolitan economies by new technologies and innovative information industries as well as through inclusive planning mechanisms and environmental assessment as a management decision process. In Chapter 7 the individual is given central focus in an exploration of the effects of personality and culture on the emergence of 'green' attitudes in corporations. Presenting case material on environmental innovations in the automobile, housing and industrial sectors, Chapter 8 examines the role of environmental technologies in providing profit and strategic improvement. In three case-studies of urban renewal and the challenges which they present to corporate innovation, Chapter 9 focuses on the symbiotic relationship between business, creative cities and sustainability. As these four chapters demonstrate, widely understood and easily learnable methods of managing environmental measurement and change are becoming available.

6 The Metropolitan Corporate Centre: Reinvention of the Medieval City-State

Reality Field 6
Metropolitan economies are being altered by new technologies and innovative information industries.

INTRODUCTION

Too often the future of organisations is explored as if they exist in a 'non-place urban realm', in the urban philosopher Mel Webber's famous phrase of the 1960s. In the technical jargon, economies and companies have become increasingly de-spatialised. They tend to live in information networks and undifferentiated airport cities spanning the globe. No matter what the country of origin, specific global icons permeate multinational companies – glass air-conditioned buildings, business suits, sleek cars, massive hotels, and corporate jets.

Yet, these icons will evolve in their own time into a new synthesis. Keynoting the new century, this pattern will have underlying environmental themes. Organisations and their communities of human beings occupy real places and real spaces. Humanity has evolved in actual places on the earth's surface, with constraining climates and limited natural resources. Within the globalised environment of the new century, there will be an intensive search for roots in real places.

Change in coming decades means the need to rediscover the 'placeness' of community and town. Even now, businesses renew regularly their commitment to ensuring local sustainable and amenable environments. These will be overarching issues in a world of globalised corporations.

In the new century, old cities, real countryside and family-based living will, for the great majority of corporate employees and entrepreneurs, provide the centre of their existence. In this globalised information economy, a rediscovery of the pre-industrial city-state with its religious, communitarian and emotional dimensions is under way.

This chapter centres on the battle in one city between the internationalist school of local real-estate developers and corporatist architects, competing against place-based group of community representatives, architects and developers.

This story of London planning is still unfolding. It typifies the problems of arriving at environmentalised businesses. As shown at King's Cross, no inherent conflicts need to be perpetuated: Planning systems can develop social housing for less advantaged groups and at the same time create world-class business centres.

THE URBANISED REVOLUTION IN CORPORATE STRUCTURES

In parallel with the need to recommit to place and heritage, the international corporate environment will be restructured. Rather than permanent staffs, temporary programme and project teams will function near the heart of international businesses. The core of international cities will become increasingly transient 'hotel' amenity landscapes, temporary offices and residential villages. In Paris, La Défense has both monuments to international companies and a changing panorama of exhibitions, tourists, and conventions. Canary Wharf and St. Katherine's Wharf in London docklands are characteristically tourist as well as business centres. New York's World Trade Centre is a beehive of smaller enterprises, as well as major companies.

The need for owned corporate accommodation will decline, as meetings are tele-networked and replaced by workshops and seminars. Urban locations will be in dedicated business centres, professional villages and corporate hotels. Corporations will increasingly have their headquarters in non-place Internet 'realms', organised as electronic networks accessible to corporate members at low cost from all points on the globe.

The urban setting will evolve from urban monoscape into a pluralistic, dense, varied urban city-state. Corporate telematics are reorganising the communications and control systems of the planet; yet the city remains a persisting centre of corporate life. It is also the centre of community regeneration.

Cities wear out. Some of their people become dependent and require systematic compassionate support. For the older industrial world, cities remain a source of wealth and competitive advantage, having at their hearts monuments, treasured landscapes, cultural memories, and educational and service establishments whose reputation reaches world-wide. For some industries – pharmaceuticals, broadcasting, electronics – at the heart of Western economic competitiveness, the universities and hospitals of the older cities comprise essential elements of their infrastructure. They produce the professionals, technicians, and intensive knowledge-development processes which are at the heart of dematerialised wealth creation.

The connection between real places and de-spatialised globalisation is one of interdependence. International business can carry out their work without cities and many are leaving the central metropolises. However, the international corporations of the Western world have in their urban institutions a matrix of cultural communication and media domination. For these reasons, in the next 50 years, more than before, issues of sustainability preoccupy urban renovation and regeneration.

For two hundred years the commercial core of cities has been the linchpin of businesses spanning continents. The skyscrapers served as giant 'filing cabinets' for people organised into large bureaucracies and more recently as central dealing rooms. Corporate headquarters have been copied around the globe, providing a centre for new instant cities: Bangkok, Djakarta, Hong Kong, Santiago, Shanghai and Cairo.

In Europe and North America, the emerging economic core city will increasingly be made up of temporary users in high-intensity environments. Surrounding them will be regenerated neighbourhoods and communities, servicing the rapidly changing core spaces. The city will become composed of interdependent communities focusing on territories and local social habitats, reinventing the organic urban models of the medieval city. Many of the skyscrapers will contain a multitude of smaller firms, rather than large staffs of single companies. All these companies are marketing to one another as well as to a global customer base.

As employees, the next generation of urban professionals may have relatively insecure economic prospects. Both urban government and corporation, two of the city's major wealth generating machines, will offer fewer permanent roles than during their heyday, the mid twentieth century. For this reason, smaller group practices and micro-consultancies will provide alternative career paths for many millions of people operating at the top level of their professions.

The need for owned business premises and corporate accommodation will decline, as increasingly team meetings are tele-networked and replaced by workshops and seminars. Corporate 'hotels' will offer rapidly changing workspace to companies with changing space needs. With each judder of environmental specification and energy costs, the demand for real estate is declining. Yet the total growth of advanced economies is increasingly focused on small firm formation and growth; many of these new enterprises will be able to pay for a position in the core metropolis, thus sustaining the viability of large cities.

Corporate headquarters buildings function as virtual environments and generate images advertising global access. In the new approach to developing urban real estate, community-based citizen participation planning procedures are increasing. The requirements of eco-planning will increasingly become the precondition of planned development at the regional level.

The implications of this process are well represented in the oldest of major world metropolises, London, as the case study of London's King's Cross demonstrates.

URBAN REGENERATION AND SUSTAINABILITY: KING'S CROSS, LONDON

In the early 1990s at King's Cross Railway Lands, London, three alternative schemes highlighted the contrast between proposals for intensive commercial development and alternative community-orientated lower-scale strategies. At King's Cross, a real battle, recognising many of the separate interests of large corporations, local low-income communities and entrepreneurs seeking to combine the needs and aspirations of both, emerged. These contending visions of the future for King's Cross, in combination with national economic retraction, opposition to the proposed rail link from the Channel Tunnel to London and lack of financing for the King's Cross project itself, resulted in a stalemate which thus far has prevented the initiation of the project, despite approval granted for the new railway in 1996 by the Channel Tunnel Rail Link Act. Nevertheless, the King's Cross case provides a rich insight into the dynamics of stakeholder interests and preferences in urban regeneration.

CASE: THE KING'S CROSS RAILWAY LANDS (KXRL)

Major infill development has been a periodic source of London's twentieth-century urban development. The redevelopment of the Barbican in the City, Thamesmead's development in southeast London, and recent air-rights redevelopment of major rail stations at Victoria and Liverpool Street suggest some of the major commercial improvements that such strategies can provide.

King's Cross has been seen for some time as a logical next candidate for regeneration. The projected completion of the Channel Tunnel in the mid-1990s was the catalyst. As early as 1985 British Rail and other land holders looked towards redevelopment, both to achieve higher returns on 'under-utilised' real-estate assets and also to finance the £2 billion cost of building a major London Channel rail terminus at King's Cross.

King's Cross Railway Lands, a land area of 134 acres/ 54.7 hectares directly adjacent to the northern edge of Central London and lying between St. Pancras Station and the King's Cross Station on Euston Road, offers similar opportunities to Liverpool Street Station to the east, completed in 1989 to general architectural and urban planning praise (Figure 6.1). However, the opportunity for King's Cross is considerably more complex due to a number of circumstances.

Firstly, the site is surrounded by, and has on it, extensive socially disadvantaged communities, including minority ethnic groups. There is a significant red-light district: sex and drugs traffic operating at the fringes of law enforcement in sordid back streets and alleyways.

Secondly, as a site for one of the principal London Channel Tunnel terminals, St. Pancras/King's Cross, the already congested area, will become even more intensively impacted during construction, lasting perhaps ten years; and thereafter. Pressure on the area was further increased by the completion in 1998 of the new British Library next to St. Pancras rail station.

Thirdly, there are strong competing pressures for provision of social housing and community facilities, as against commercial and regional accommodation. There is a significant argument against any form of further development which can only intensify congestion. However, extensive dilapidation and under-investment, linked with many buildings of historic character, suggested that regeneration was desirable.

HISTORY OF THE SCHEME AT KING'S CROSS

The London Regeneration Consortium (LRC) proposals

King's Cross Railway Lands first emerged as a scheme from the major landowner, British Rail, in the mid-1980s. In October 1987 British

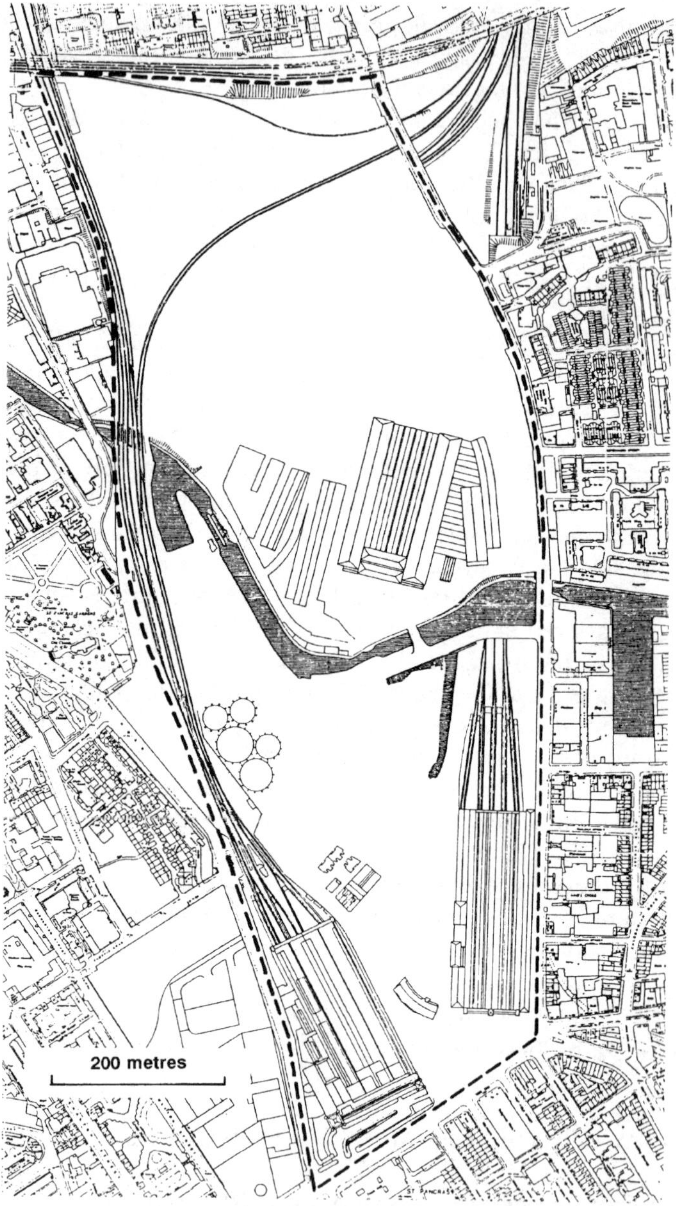

Figure 6.1 Map of King's Cross Railway Lands (north at top of figure)
Source: Ledgerwood *et al.*, 'Environmental Assessment as a Real Estate Management Protocol', *Cities*, vol.II, no.3 (1994).

Rail (BR) prepared a development brief for the site and invited four developers to submit proposals to meet the requirement of the brief. By January 1988 the developers had been reduced to two and the schemes were put on public exhibition. One development team comprised the recently formed London Regeneration Consortium (LRC), a conglomerate comprising the British Rail Property Board, National Freight Corporation and the developers, Rosehaugh & Stanhope. This team put forward two alternative schemes, one proposed by Skidmore, Owens and Merrill and the other by Norman Foster Associates. The rival consortium or development team comprised contractors, McAlpine with developers Speyhawk and their architects, York, Rosenberg and Mardell. In June 1988 BR selected the LRC proposals designed by Norman Foster Associates as the preferred option (Figure 6.2).

The developers had achieved financial success through association with two previous British Rail terminal schemes, Broadgate at Liverpool Street station in the city of London and Charing Cross station. The linking of successful large-scale developers, with innovative urban design achievements already built or under way, and coupled with a British architect of international stature, was initially seen as a formidable, if not unstoppable, force for development of the area.

Furthermore, the year 1988 was the apex of London's property boom; thus, the winning scheme reflected standards which already had achieved architectural praise and financial success. The LRC proposals focused on commercial buildings, a huge concentration of large office blocks, intermixed with what appeared to be largely upmarket apartment blocks. A major urban park was to be created in a central open space to be opened up by removal of existing buildings, including a number of important historical interest. At the north end of the site, two skyscraper towers of fifty storeys each were to create a trophy summit. The LRC proposals included residential accommodation of varying tenures and sizes as well as substantial commercial development. High-quality office facilities were to be provided to meet anticipated national and international demand for London office space. Traffic proposals included improved pedestrian access achieved by closing Goods Way, a road which bisected the site and which carried heavy goods vehicles, in an attempt to improve the local environment.

As a part of the London Regeneration Consortium, British Rail were also designing a major third rail terminus at the southern end of the site, to receive Euro Tunnel Rail traffic from the late 1990s.

Figure 6.2 London Regeneration Consortium proposals
Source: Ledgerwood *et al*., 'Environmental Assessment as a Real Estate Management Protocol', *Cities*, vol.II, no.3 (1994).

Development on the whole of the site was expected to contribute to the financing of this terminus. BR proposed changes to the railway layout including new low-level connections between the new international railway line and the existing midland and east coast main lines. The new international station would also have included provision for suburban connections at low level. The BR proposals, which were originally developed as part of the LRC proposal, were subsequently abandoned along with the LRC plan.

Public reaction was at first muted and then, to the surprise of the council and the developers, infuriated. Clearly part of the strategy involved removing from King's Cross some of the social and economic communities living and working around the site. The office skyscrapers would be built adjoining carefully restored Victorian urban neighbourhoods, which housed a wide mix of middle and lower income households in adjoining conservation areas. Fears that the evolving pattern of social land uses and affordable housing resources would be destroyed by a development on this scale, particularly with the proposed high-rise fixation, led to strong public criticism.

As a consequence of this reaction, in June 1988 Camden Council, the local planning authority, produced its own planning brief for the site which emphasised the need for local employment opportunities. A minimum of 1,850 social housing units was specified, together with a range of other social and community uses. The brief did not stipulate any maximum levels of commercial development for the site nor did it specify any particular mix or distribution of land uses.

All proposals for redevelopment of the King's Cross lands require an Environmental Statement to fulfil requirements of the United Kingdom Regulations, Statutory Instrument 1199 (1988), Assessment of Environmental Effects.[1] This statute came into effect as the competition winner was announced. Thus, although technically avoidable, it was felt that the environmental assessment should be carried out to ensure that full safety, health and environmental considerations (SHE in current jargon) were being followed. The regulation embodied the European Community directive on Environmental Assessment of major projects of the same year.

LRC's advisors on environmental impact assessment (EIA) were the surveyors, Montagu Evans. The actual environmental assessment on the LRC proposals was carried out by a specialist firm called Environmental Resources Ltd (ERL), the UK's best-known environmental science and assessment practice. As was the convention in UK practice, the EIA focused solely on physical elements and in particular soil contamination issues. However, in the eyes of many, the greatest relevant environmental impacts were the social and economic effects of the proposed development which had been muted.

Development of a community planning counter-strategy: the 'Planning for Real' episode, 1989–91

By this time other influential actors in Camden local communities were involved in arguing the future of King's Cross Railway Lands. A number of local councillors were chairing local residents' groups and by 1990 two groups, the King's Cross Team and the King's Cross Railway Lands Group, had developed proposals.

King's Cross Railway Lands Group (KXRLG)

In the highly politicised atmosphere of Camden Council in London, local groups, in particular a community organisation called the King's Cross Railway Lands Group, had the support and attention of Labour councillors and party leaders. By 1991, alternative proposals were being developed by KXRLG which resulted in an option which, although based on KXT proposals, included the important amendment of removing all commercial office space in favour of additional high-density social housing. Because this plan was specified not to be commercially viable, for the purposes of this analysis it is not given further consideration.

King's Cross Team (KXT)

The KX Team was formed in 1990 through the financial sponsorship and intervention of a local investment group, led by Martin Clarke. His motives were both commercial ('leadership of a workable alternative') and philanthropic ('contribution to community improvement'). The stated aim of KXT was to produce a King's Cross lands strategy which, while being commercially feasible, also responded to community needs. KXT's brief was to develop an alternative to the LRC proposals which would bridge the gap between the investment needs of the landowners, on the one hand, and the social–environmental needs of community groups. The KXT proposals, with an accompanying Environmental Statement, were published with financial help from London Electricity plc in 1991. The Preliminary Environmental Statement reflects the existing known situation about site conditions as published in documents in the public domain, supplemented by technical documents. The KXT proposal was for a complex commercial and residential development, covering 134 acres (54 hectares), and would have affected two Conservation Areas, one along the Regent's Canal running west to east through the site and the other around King's Cross and St. Pancras Rail stations along the southern edge of the site. (See Figure 6.3.)

Figure 6.3 KXT Proposals

Source: Ledgerwood *et al.*, 'Environmental Assessment as a Real Estate Management Protocol', *Cities*, vol.II, no.3 (1994).

In addition to developing many original urban design analyses, maps and architectural proposals, KXT used published data for proposals of London Regeneration Consortium (LRC).[2] In the environmental assessment carried out by KXT, these sources were supplemented by site inspections and interviews with designers, social workers, planners, and local representatives.[3] Norman Sheppard and Ian Haywood led and participated in a community design exercise covering 13 months in 1990–91, assisted by the social planner, Michael Parkes.[4] A series of Planning for Real weekends involving local councillors, residents, and planners were held at University College London in summer and autumn 1990[5] (Figures 6.4 and 6.5).

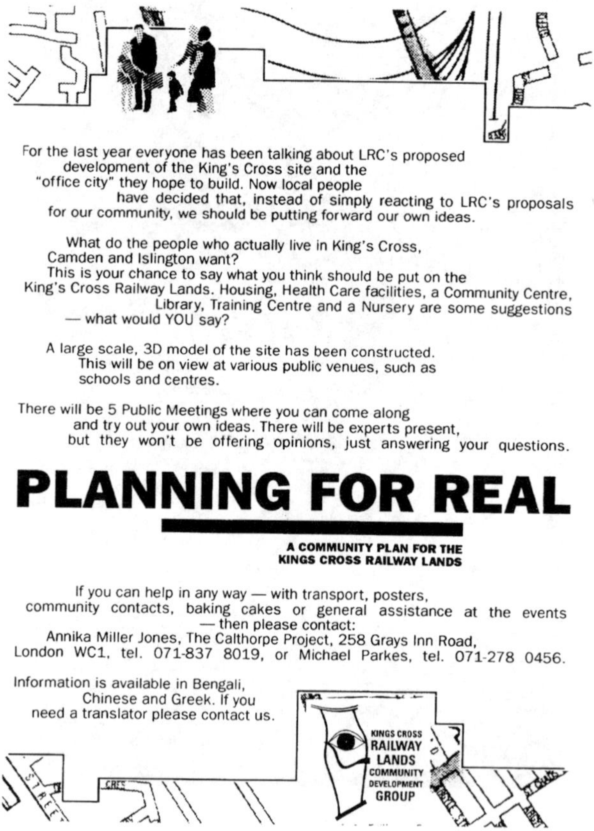

Figure 6.4 KXT 'Planning for Real' announcement

Source: Ledgerwood *et al.*, 'Environmental Assessment as a Real Estate Management Protocol', *Cities*, vol.II, no.3 (1994).

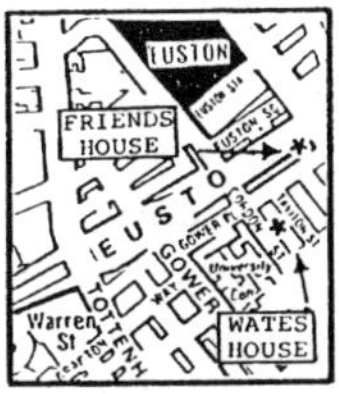

KING'S CROSS PLANNING WEEKEND

KXT

THE ALTERNATIVE APPROACH

Thursday 13 September to Monday 17 September 1990

You are invited to participate in this Planning Weekend organised by KXT, a team of architects and planners commissioned by Martin Clarke to produce an alternative scheme for the 134 acre King's Cross Railway Lands Site. This Weekend will make use of the joint Railway Lands Group and KXT 'Planning for Real' exercise.
The purpose of the Weekend will be to produce realistic proposals that will balance the needs and aspirations of the local community with those of London and the wider region.
Key speakers will present for public discussion a series of local and London issues. They will be supported by an expert Design Team who will participate in a series of workshops during the Thursday to Saturday public sessions. On Sunday and Monday the Design Team will prepare draft proposals ready for presentation to the public on Monday evening.
The Planning Weekend is a key event where you will be able to help brief the Design Team and influence the form of 'THE ALTERNATIVE APPROACH' for the largest development site in London.

THE BARTLETT SCHOOL OF ARCHITECTURE
Wates House • 22 Gordon Street • London WC1

THURSDAY 13 SEPT

7.30 - 9.30 pm **Public Meeting**
- Introduction and background to Weekend
- Report on 'Planning for Real'

Open forum

FRIDAY 14 SEPT

10.00 -11.00 am **Session 1**
History of the Area
- History of King's Cross
- Conservation areas
- Existing uses and users

Speakers to include:
Michael Hunter, Civic Society
Malcolm Tucker, GLIAS
KXRLCDG
Open forum

11.30 - 1.00 pm **Session 2**
Planning Context
- Regional
- Local - Camden & Islington
- Community overview

Speakers to include:
Giles Dolphin, LPAC
David Pike, Camden
Ian Crawley, Islington
KXRLCDG
Open forum

2.00 - 4.00 pm **Session 3**
Local Needs
- Housing & special needs
- Employment & training
- Community development
- Retail & local services
- Arts, leisure, recreation & tourism

Guest speakers invited
Open forum

SATURDAY 15 SEPT

10.00 - 11.30 am **Session 1**
Sustainable Cities
- Sustainable cities
- Healthy cities
- Economics
- Energy

Speakers to include:
Michael Edwards, UCL
Pat O'Sullivan, Bartlett
Open forum

12.00 - 1.30 pm **Session 2**
People, Process, Product
- Community development
- Quality & the workplace
- City living
- Moving people
- Environment

Speakers to include:
Jon Aldenton, Tower Hamlets
Trevor Hendy, Chapman Hendy
Adrian Leaman, Building Use Studies
Open forum

2.30 - 4.00 pm **Session 3**
Workshops - Local Needs

4.30 pm **Workshops Feedback**

MONDAY 17 SEPT

8.00 - 9.30 pm **Public Meeting**
Friends Meeting House
Euston Road, London NW1
- Report back on Weekend
- Recommendations & proposals

Open forum

For further information please contact : Berri Shulman-Hill
071 278 0456
Beverley Davies
071 485 8555

PLEASE COME AND PARTICIPATE. EVERYONE WELCOME. REFRESHMENTS PROVIDED.

KINGS X TEAM • P O BOX 1978 • LONDON NW1 1UQ • TEL: 071 278 0456

Figure 6.5 KXT 'Planning Weekend' announcement

Source: Ledgerwood *et al.*, 'Environmental Assessment as a Real Estate Management Protocol', *Cities*, vol.II, no.3 (1994).

'Planning for Real' is a recent British approach to community participation in strategic planning decisions but the idea has its origins in much earlier work both in Europe and in North America. In essence, through a series of open preparatory meetings and reports, community groups living on or near a major development are asked to prepare their 'wish lists', on the assumption that the physical environment will inevitably be undergoing change. Where possible, drawings, sketches, mockups and cartoon murals of their ideal desires for the development are prepared. Sometimes local planning, architectural and art colleges are able to lend students and staff to assist. Participating groups may include school classes, nursery groups, pensioners' groups, community associations, conservation groups, and neighbourhood watch groups, employer groups, commercial associations, professional and trade associations – any group identifying with a particular area where environmental change is proposed.

After each group has prepared its own report and communications, an event, or series of events, is brought together by environmental professionals assisted by community development leaders. Groups meet and negotiate with one another. Press and media are invited and a record of the event/s is prepared and published and/or broadcast. Under the guidance of professionals, political leaders and responding to specific inputs from outside 'experts', further consensus-based strategic views of development are agreed. Finally, at the end of a process covering days or weeks, the entire preparation is summarised in a single document which should report within it an indication of every participating group's contribution.

As a result of the process, environmental planners are then entrusted to prepare a single planning report, condensing one scenario, which is then the basis for preparation of a formal plan.

The politics of the planning process: two proposals

Although the ideal process produces a single plan, the process conducted at King's Cross during 1990 resulted in two plans: the KXT Plan reported here and the second plan developed by the King's Cross Railway Lands Group (KXRLG). The Planning for Real process, covering a year, was reported in the two published plans and in a Channel 4 documentary broadcast in November 1991.

The problem faced by Camden Council, however, was formidable. Not only were there now several proposals, but by 1991 a frequently quoted level of expenditure on the LRC scheme was £30 million for the studies alone. Moreover, the major landowner, British Rail, favoured large-scale commercial development which would result in cashflows adequate to finance the projected £2 billion Channel Tunnel terminus at King's Cross. Although Camden had been in extensive consultations with LRC, the changes to the proposals accepted by LRC were marginal. Camden Council wanted to see the site developed, but were faced with substantial local opposition; hence the decision to solicit plans from both the London Regeneration Consortium and the King's Cross Team.

Comparing and contrasting the KXT and LRC schemes

The 1992 published plans of KXT and LRC present a series of detailed contrasts, as well as radically differing perspectives on the future of London's development. On the one hand LRC focused on international trophy architecture in an overwhelmingly commercial new London district, resembling the environs of Regents Park a mile to the west. KXT centred on meeting the need for both social housing and a smaller proportion of commercially viable offices and shops. Table 6.1 contrasts these two schemes with respect to key provisions.

LRC proposed a commercial enclave such as the successful railway lands development at Broadgate, Liverpool Street Station in the City of London. The KXT strategy, instead, expanded the concept of a London urban quarter, combining both local and metropolitan functions in an environmental setting sympathetic to existing heritage values around King's Cross and the Regent's Canal.

Two broad objectives were built into the KXT proposals, namely (a) a substantial provision of commercial and metropolitan development around the rail stations, and (b) a residential area to the north with a range of community and social housing requirements. LRC indicated a minimum of 60 per cent social housing, which was allocated to less than a quarter of the site. The following comparison of the two schemes will address environmental categories, housing and community benefits, metropolitan land uses, ecological impact, financial gains, open space and impact on real estate development policy.

The environmental assessment procedure in the two schemes

LRC in 1988 commissioned a leading consultant, Environmental Resources Ltd, to carry out an environmental assessment. ERL

Table 6.1 KXT and LRC proposals: summary comparison

Provisions	*KXT*	*LRC, 1991 amendments*
Great Northern Hotel, listed Grade II	Retained and integrated into new concourse	Demolition
Other listed buildings	Retention & conservation	80% Demolition
Ecological park	Retain and expand	Submerged in new commercial development
Maximum park size	1.0 acres	34 acres
Social housing units	1750 units	650 units
Family housing	60%	50%
Social housing proportion	66%	46%
Commercial office space	4 million sq. ft.	5.8 million sq. ft.
Space for cultural groups	Key feature	Absent
Community health facilities	Key feature	Absent
Space for low-tech industries and small business	Key feature	Absent
Energy conservation, waste recycling and low-cost energy resourcing	Central to the design strategy	Unknown from published plans
Wildlife and nature conservation, heritage design & open space	Key feature	Absent
Return on investment	26%	55%
Defensible open spaces	Key features, small piazzas	Open spaces are large and indefensible
Design guideline	Key feature: sound environmental and ecological principles	Absent
Bus links	Maintained across site	No provision
Safe pedestrian and cycle routes	Maintained throughout site	No provision

utilised the categories for assessment identified in Table 6.2, but omitted issues of social and economic impact. This omission conforms to widely established UK practice as defined by statute. It differs, however, from longer-established North American protocols where socio-economic impacts have for some time been seen as essential components of the planning process.

The KXT assessment procedure, on the other hand, studied primarily the social and economic impacts. Both assessments as published comprise a review of differing approaches to large-scale development, rather than of final components of detailed building design. Because its assessment was completed later and with the benefit of very considerable investment in site research and assessment already published by LRC and by Camden Council, KXT was able to include more categories and socio-economic impacts as well as to establish the Baseline Information about the site. This information

Table 6.2 Categories reviewed in the environmental assessment

Categories	*KXT*	*LRC*
Air quality	+	−
Composite physical constraints of site	+	+
Construction phase: displacement impacts	+	−
Ecology	+	+
Ecological sustainability	+	−
Economic sustainability	+	−
Financial/commercial stability of local business	+	−
Financial programming	+	+
Financial robustness	+	−
Financial sustainability	+	+
Heritage and conservation	+	+
Listed buildings	+	+
Microclimatic planning	+	−
Noise	+	+
Pedestrian walkways	+	−
Physical environment: existing site conditions	+	+
Socio–economic assessment	+	−
Soil	+	+
Street furniture and environmental fittings	+	−
Traffic	+	+
Urban design & landscape characteristic	+	+
Vibration	+	+
Views	+	+
Water	+	+

was used to ensure that land uses are proposed in the most suitable locations. Impacts of the proposed construction and development on the environment were considered in the KXT Environmental Statement, as well as mitigating strategies to offset negative impacts.

Housing and community benefits – the northern reaches of the site

In KXT's northern 'King's Cross New Town', development will link to the older community through continuity in scale and use from surrounding streets into the new development. Because the urban quarter will take shape in a less intrusive manner as it expands into the community, it will be more easily integrated to existing areas. The Town will provide on-site social housing in a range of sizes, tenures and cultures. This strategy would consider best allocation of existing and proposed housing stock in order to provide improved capacity for both family and smaller-household units.

By contrast, LRC proposals constitute a massive land-use change, essentially wiping out the viability of many neighbouring uses because of initial construction damage and final land-value increases.

Family housing

Family housing may be better concentrated in larger units characteristic of older residential neighbourhoods near the site, while the new development may be most efficiently used for housing for smaller households. Households with 1 or 2 persons comprise a majority of inner London occupancies. In any case, a need for substantial family housing accommodation in nearby communities is well documented, but ideally may be best provided in existing neighbourhood settings.

In both schemes, family housing on the site is, given pre-existing industrial contamination in many locations, less than ideal. Trace elements of heavy metals – lead, cadmium and other toxic substances – persist in house gardens, even after decontamination. Long-term subsoil deposits reflect industrial uses on the site over the last 100 years.

Social housing impacts – the main asset of KXT

Within the KXT scenario, the development's impact on social housing availability, over the present baseline, will be to increase the supply of new social housing. Its impacts on overall rentals and markets cannot be readily separated from those characteristic of change and expansion of Central London land uses into the areas around King's Cross. However, there is a greater likelihood that the nature of the proposals will ease such pressures, in two ways:

- The site's housing stock itself will be predominantly in Housing Association and LA housing rented tenure; it will help re-house some of the 2,000 households currently homeless in Camden.
- It will considerably expand the total local housing stock available at affordable rents and thus ease pressure on the remainder of the local housing stock, thereby restraining rent increases.

LRC's housing mix would have instead placed much greater numbers of up-market apartment dwelling blocks in the areas immediately north of the main rail stations. Substantial but smaller amounts of social housing would have been permitted in the northerly extreme of the site, but at much greater density than nearby neighbourhoods.

Metropolitan land uses

In both schemes, at the southern portion of the site, up to the Regent's Canal, a range of developments catering to the national and international development opportunities are scheduled. In the KXT scheme, these will be developed in a manner which will respect local scale and heritage in the Conservation Area. Under LRC plans, the priority was commercial viability and, as a result, many existing heritage buildings would be demolished.

Ecological impacts

Under the KXT approach, the physical standards and objectives being designed in this Strategy are consistent with the new green development standards of the last decade of the twentieth century. Multiple provisions for nature conservation are built into the proposals.

By contrast, LRC gave little or no consideration to adopting green development standards for either the commercial or residential developments.

Trade-offs between financial and social gains

LRC's approach, at the requirement of British Rail and as accepted by Camden council, was to maximise financial returns and meet the Council's objections by offering site benefits. KXT proposed, in response, a lowering of financial return to the level necessary to meet a range of social, economic and environmental objectives.

The cost of this on-site balancing of community-linked objectives is the loss of some of the large financial surpluses of a high-density commercial development scenario. Thus, in the KXT strategy community gains are advanced alongside significant commercial gains. Larger financial gains anticipated in the LRC 1988 proposals looked unattainable by 1993 as the London office market was glutted with new unoccupied commercial and retail buildings. Any off-site benefits would have evaporated.

In terms of financial feasibility, the KXT proposal suggests a level of low-scale phasing of development based on a balance of uses which can tolerate wide fluctuations in the London property market. It thereby has a greater likelihood of medium-to long-term financial viability and completion.

Open space proposals

Under KXT proposals, the proposed development will provide small, safe, better secured and richly patterned open-space provision in series of networked spaces which can be controlled by the community and made resistant to delinquency cultures. These will largely be in the form of enclosed inner-block gardens specific to each secured block of housing.

LRC's focus on a single large open space and contributory green avenues has the benefit of enhancing land values in the northern part of the site. These spaces appears to invite re-colonisation of the area by presently well-entrenched drug-dealing cultures. It is conceivable, however, that a walled and well-secured series of open spaces can, on the pattern of Belgravia and Bloomsbury, be retained as socially viable, but at substantial security costs.

KXT's development strategy makes possible, through a programme of interim cross subsidies, the continuation of many existing businesses on the site which would be relocated in phase with the construction programme. Their final destination would be in specified areas on the site.

During construction, substantial disruption of existing businesses and of present housing is inevitable. In order to reduce construction and displacement impacts on residents and businesses to a level which can respect local needs and community and Council priorities, codes governing construction, including hours of movement, dust and noise generation, are needed.

No doubt due to commercial priorities of a client-driven brief, the LRC brief gives little consideration to the need for replacement accommodation for local businesses displaced by construction.

Real estate policy: implications and impact on design

Recommendations were put forward by KXT to be considered as part of conditions for planning consent on the developer's Outline Planning Application. Recommendations have real-estate policy and urban development policy implications and are summarised below.

1. Establishment of research leading to a *Design Guideline and Environmental Audit Procedure* to cover design programming and per-

formance specification on new and renovated buildings on all site locations.
Implication: Planning permission can be used to ensure a high standard of environmental assessment and audit across the site as *a* whole and throughout the development process.

2. Adoption of *Good Contractor Code and Guidelines* concerning construction phase.
Implication: The construction process should be as rigorously tested for and controlled by Environmental Impact Assessment (EIA) as the final output, particularly in large projects in regeneration settings.
3. Implementation of a *pre-construction research programme relating to soils and subsoil conditions* throughout the site, in areas which will be excavated and built through, to ensure that health and safety hazards in surface and subsoil deposits are precisely measured and removed prior to, during and after construction.
Implication: Environmental assessment can be used to establish and enforce careful environmental health procedures.
4. Consideration of *Joint Housing Strategy* involving proposed social housing complexes on this site and in adjacent neighbourhoods in the Boroughs of Camden and Islington.
Implication: EIA can be used to focus back on larger strategic issues affecting other parts of a statutory LPA process.
5. Establishment of a King's Cross Lands *Local Business and Employment Strategy* to ensure integrated and sensitive provision of new premises, relocation phasing and business and job training programmes.
Implication: EIA is as much about economic development in a sustainable manner as it is about environmental planning.
6. The formation of a *Community Development Trust*, as a partnership between the community, developer, investor institutions and local authority to oversee the development.
Implication: Environmental assessment can lead to appropriate environmental management structures using resources of relevant stakeholders. The outcome here emphasises that less intensive development scenarios offer less risk.

As in other complex changes in the urban environment of the information society, urban regeneration can be made part of sustainable and socially aware development. Furthermore, commercially aggressive alternatives become less viable if they are subjected to full-range commercial-risk assessment. This appraisal is part of the socio-technical systems around environmental impact assessment.

Environmental impact assessment as a management decision protocol

The management implications for real-estate policy is that Environmental Impact Assessment (EIA) is a tool which in its application to a large site can be turned to many uses both by promoters and by statutory planning authorities. Most centrally, it can be used to focus discussion on the contending interests and priorities, inevitable in major projects. In the case examined, LRC's original brief clearly was drawn to ensure that many local interests were deliberately unexamined and thus omitted. The weight of expertise, financial backing, previous success, and international pressure would drive the massive proposal through on sheer momentum. With the property crash, together with increased uncertainty about possible funding of the international rail link and the substantial financial re-juggling faced by the member companies of the LRC consortium, this momentum was lost.

The complex nature of large urban projects, with their numerous stakeholders and competing interests, requires a decision process which has built into it the potential for multiple contributions. These contributions must reflect a range of socio-cultural readings and inflections, reflecting the many subcultures normal in dense metropolitan settings in the twenty-first century.

Because they pay attention to many details as well as to strategic urban design, environmental assessment offers such a tool. As a statutory procedure it can be carried out as a minimal qualifying feature, as has been done with LRC's 1988–91 proposals. Alternatively, as was attempted in the KXT project, assessment can be used to underscore and highlight the different payoffs which alternative design scenarios indicate.

Thus used, EIA is political. It allows planners and residents to consider in detail the different implications, however conjectural, for their community of contrasting proposals for change. It also allows the chance for detailed trade-off negotiations – 'planning gain', in the now unfashionable parlance – between developer and local planning authority.

In 1993 there was little indication that either of the local councils or the developer were negotiating on the basis of the published assessments of either LRC or KXT proposals. However, it increasingly appears that the LRC proposals will founder because two of the key precepts of the scheme, a buoyant commercial market and government commitment to a low-level international rail link disappeared. Failure to develop proposals with a high degree of physical and financial flexibility, coupled with the failure to identify, through any form of assessment, the crucial reliance of the scheme on these linkages means the proposals are fatally flawed.

During this period it is likely that the relative merits of various proposals, as seen through the environmental assessment procedure, will no doubt be reappraised and become more self-evident. Properly deployed, the EIA protocol may have the effect of moving the property development industry into new ways of designing and regenerating major urban complexes. It provides a formal and legal mechanism by which such concepts as planning balance sheets can be adopted.[6]

Essentially it is a tool which suggests for the industry a new ethos, a culture of values, and a system of ethics, prime requirements for social contribution. EIA provides the opportunity to understand better linkages among different components, costs and benefits of proposals for urban regeneration. Its procedures can help decision makers compare alternative strategies for complex urban regeneration schemes. Because it is a recent approach to planning new environments, EIA is becoming a key management protocol in deciding strategic and social issues in urban design and regeneration.[7] In general, EIA as the case examined here shows, in complex urban settings, socio-economic and cultural assessment of environmental designs and proposals generates important, if more subjective, understandings of urban change.

EIA, in its development as a management protocol, is important to business theory and practice for three reasons. Firstly, it brings potential social, economic and environmental conflicts into focus at a much earlier stage of the investment procedure. Thus, promoters and other stakeholders supporting a decision must specifically identify potential costs which balance purported planning benefits. A possible consequence is that early planning stages can achieve a much greater level of design responsiveness for impacted users and neighbours.

Secondly, environmental assessment is now required prior to outline planning application, through a Europe-wide statute – the 1988 Directive requiring environmental assessment of major projects. Costs of consulting with potential objectors are borne by promoters rather than the statutory local planning authority. This should lead to both a more socially responsible real-estate development industry and a more efficient use of public planning resources.

Thirdly, potential financial backers for the first time will have early signals of the difficulties faced by real-estate schemes they are investing in. This early-warning aspect may make investors conflict-adverse. Some investors may become less ready to gamble on substantial gains made less certain by assessment. Moreover, by careful notation and publication of costs and benefits of development on community and environment through EIA protocol, these gains are less easily captured.

As a consequence EIA should produce a new real-estate development paradigm over coming years. Emphasis will be on careful design and pre-marketing to affected users, investors, and stakeholders. Traditionally associated with a piratical reputation, the property development industry may find itself evolving new business systems. As a key actor in the formation of cities, the industry will become more systematic, knowledge-led, and inevitably less secretive. There will be increased likelihood of generating socially effective urban complexes where community gains are measured and valued alongside financial returns.

Since the Second World War, the world property industry has offered remarkable financial returns, as well as, at the end of the 1980s, some huge losses. The volatility of property markets allows for key strategic opportunities to investment groups – pension funds, insurance companies and other large international institutions. As the new systems of environmental assessment will make explicit many of the spillover costs and potential returns, the attractiveness of property as a repository for long-term investment may be altered.[8]

Environmental assessment: a definition within the terms of decision protocol

Decision protocol, as an aspect of decision theory, relates to the need for agreed procedures for resolving disputes within named decision contexts.[9] Environmental Assessment introduces a new layer of decision analysis and option appraisal into urban strategy.

One basis for contemporary decision theory are the sub-optimising ('satisficing') models of Simon. In the Simonian paradigm for decision-making, complex organisations have neither time nor resources to explore all options in a major investment choice.[10] Thus, search and appraisal operations continue only until such time that the organisation is required to choose and has found a best-approximation to an ideal. The ideal may itself be implicit and lacking in consensus.

Thus, in planning large urban projects, there is in general a limited range of exploration of options. Indeed, the planning authority may depend on private-sector entrepreneurs to offer a single alternative, which is then either accepted – 'given planning consent' – or denied. If denied, there are then various level of appeal to government.

Environmental assessment adds considerable power to public decision-taking in this procedure. Essentially assessment requires that the corporation, rather than government, take responsibility for first discovering and then costing the environmental consequences of development. Importantly the construction period, which in major projects stretches over years, is also required to be assessed for its environmental impacts.

In Simonian terms, therefore, the government is seen to be increasing its ability to advance towards optimality. Whereas prior to EIA, consent may be seen to provide, for example, 30 per cent of optimality, it may be that EIA adds a component of information that can advance the quality of knowledge by a useful quantum – notionally, perhaps another 15 per cent, or a total of 45 per cent of a conceivable, but indescribable, ideal. In urban planning procedures, EIA, therefore, introduces more information earlier into the decision process.

CYBER-METROPOLIS AND THE RISE OF URBAN TELEVILLAGES

In the discussion of development of metropolitan centres, the focus has heretofore been on the United Kingdom and the proposals for regeneration in central London. Developing countries are also experiencing rapid urban growth and devising complex strategies for the future. The emerging cyber-metropolis at Kuala Lumpur, Malaysia, illustrates the extent to which some developing countries are leapfrogging first-world technology into the third millennium.

A model for the new century's globalised cityscape

Despite the economic setbacks in Southeast Asia in the mid-1990s, nowhere is the stark challenge of the developing tiger economies more clearly articulated than in Malaysia. In the late 1990s Malaysia was putting together an environmental, urban and technological strategy which could challenge any holistic innovation setting on the globe.[11] The strategy was based on a decade of economic growth at 8 per cent-plus each year. In spite of a population where many people were still living on subsistence agriculture, Malaysia had achieved the status of the world's largest producer of air conditioners and video recorders. As late as 1975, its export economy had been dependent on palm oil, rubber and tin. Now, as well as many technology products, it exported the most demanding of technological products, cars, which competed effectively in Europe and North America.

More importantly, its government had taken a positive stance to capture the future, rather than fight for contemporary techno-industrial supremacy. Its approach was rivetingly novel. The first part of this approach, developing a major capital city, was in place by mid-decade. It developed its capital, Kuala Lumpur, as a major regional metropolis. In 1997 the tallest buildings in the world, the Twin Towers, were completed. These Towers, with their pointed tops and enormous stainless-steel-clad profiles, seemed to be the epitome of space-age architecture, providing a confident counterpoint to the skyscraper revolution which had begun in Chicago a century earlier.

Putrajaya – a Cyber Metropolis

Twenty miles to the south, at Sepang International Airport, was the second strand of its international strategy to confirm its role as a major metropolitan centre for the whole of South and East Asia. To connect Sepang and Kuala Lumpur, finally, a third major element was being designed with the assistance of the world's largest companies, including IBM and MicroSoft – an 800-square-mile technological metropolis linking the two main points of development, named Putrajaya.

This Cyber City is planned as a comprehensively wired living and working city, with a third of its land area remaining in ecological reserves, including native jungle habitat and in some cases tea plantations. Many of its constituent industries will be in the new services sectors. For instance, there will be an extensive medical and biotechnological industry. Practitioners will offer a fusion of Avuryedic, Chinese and Western medicine. Most dramatically there will be extensive 'telemedical' facilities, with resident experts able to advise medical colleagues via the Internet on 'real-time' diagnostic and treatment requirements.

From within the Cyber Metropolis, national government and public services are also being digitised. Every citizen will carry a digital identity card which will be keyed into computers of government's social ministries. Thus, computers will take care of such routine tasks as issuing driver's licences and paying pensions. Such wholesale systematisation of public services may be resisted in some countries with high levels of individualism in their political tradition. However, within the Confucian traditions of East Asia, influential in Malaysia, such 'communitarianisation' of state functions leaves, as Malaysia's leaders assert, more time for civil servants to provide more real face-to-face services. This will be the first government on the planet where every citizen will carry this multifunctional smart card, which will enable direct access to government, banking, credit, telephone, transport and club services.

In a similar manner, the new homes and schools throughout the community will drive electronic networks through all existing learning and domestic information systems. Schooling will be divided into 'left-brained' tasks which computers will assist through educational learning packages – fact-based subjects, analytic skills as in maths and chemistry, language learning. This will free teachers for 'right-brained' learning – involving creativity, design, synthesis and team problem-solving.

Putrajayang will become a global island within the larger nation of Malaysia. Its government, called the Multimedia Development Corporation, will have its own law systems. There will be the unrestricted right to recruit workers from all over the world. Freedom of ownership and lack of censorship, including Internet, will be established and be distinct from the remainder of the country. The target is to attract the most innovative new businesses on earth and the professionals to man them. To enable this effort, already specific laws governing digital signatures, digital contracts, and digital intellectual property protection have been established.

The cyber-residential environment

Asif Nun, chief planner of Putrajaya, describes residential areas as 'cyber villages'. Every home is connected via optic fibre to the Internet. The overall connection features a 10 gigabit digital highway with direct connections to other ASEAN (Association of Southeast Asian Nations) countries, North America and Europe. Local shopping and commercial centres will offer goods and food from all over the world: 'Asianfusion' food will become a major export industry in its own right.

The corridor as an environment will inevitably destroy some areas of jungle and tea plantation. However, a third of the land area will be maintained in green parks and reserves and there will be a low-rise – maximum 5 storeys – development policy intended to sustain the intimacy and quality of traditional Malaysian towns and villages.

Typically a manufacturing company will bring together components produced in China and Japan, with Indian-designed software, and be financed from Malaysia's own offshore financial services island, Labua.

Implementation

The realisation of this vision in the first decade of the new century will require a huge national effort in Malaysia, where existing education and industry are only beginning to approach those of, for instance, Eastern Europe. To bring about CyberMetropolis, the national government has recruited an international Board of global business leaders, including Bill Gates of Microsoft and Louis Gerstner of IBM. Its first board meeting was held at Stanford University in California. As Martin Jacques comments, 'If this latest gamble with modernity pays off . . . then the ideas that inform it will be progressively applied across the country. Malaysia will have moved from a rural economy to the information age with little more than a hop, step and a jump.'

More soberingly, once established, this 'new paradigm' of national development will challenge every country in the coming century, in terms of products, services and efficiency in development. Its potential success will alter the future everywhere, just as European and American development models of a hundred years ago altered the rest of the planet.

Enterprise televillages in advanced economies

Opportunities arising from the digital economic revolution will have even greater effects in advanced industrial economies' cities. The changing cityscapes of revitalised urban regions have many older neighbourhoods and districts with worn-out buildings and semi-derelict land uses. These brownfield sites are producing a range of habitat solutions for small businesses and young enterprises which in research at the University of Greenwich Business School have been brought together under the classification of 'enterprise televillages'.[12]

Centering on processes of local innovation, televillages will offer many choices to new entrepreneurs, whether in early career or "third age "business startups. It is predicted that by the third decade of the new century, many different versions of these regenerated incubators of community and prosperity will exist around the world. Distinctively they will be locally based communities based around microbusinesses – 1 or 2 person enterprises – marketing to global marketplaces.

Televillages, as well, describes the upscale mountaintop retreats of silicon billionaires in Colorado, Montana and the Swiss Alps. Because of their relative distance from metropolitan centres as well as their dependence on small numbers of the super affluent, they cannot be seen as general models for the televillage. The most general approach will be that reflecting the desire and need for digitally based smaller businesses to cluster together and to offer a reasonable support to its members, many of them younger and certainly a number of older households.

Two important British prototypes stand out in 1998 research conducted under Housing Corporation funding at the University of Greenwich Business School. The first is in a rural area in North Wales, the second at the urban heart of the Newcastle upon Tyne, a major city in northeastern England. Under entrepreneurial leadership of Ashley Dobbs, the Crickhowell Televillage since the 1980s has created a expressly designed and purpose-built arc of terraced houses at the edge of the village. Each house is owned by its occupants and in 18 of the 20 units in 1996, a small business, started by the occupants and based on telecomputing, was operating.

At the heart of Newcastle's inner-city regeneration district near the main train terminus, the Northeastern Newstart Centre has since the late 1980s been exploring how innovative computer-based businesses can be cultured within older urban structures. Using a range of disused and low-valued shops and office blocks, small enterprises have developed following renovation and ISDN wiring. In 1998 there were some 40 businesses, many with residential premises connected or nearby, in this complex.

As noted in an earlier book (*Greening the Boardroom*, pp. 115–18): 'The concept centres around residential apartments with attached offices for entrepreneurs working from home. The numbers involved may be as few as two and as many as 100 units in each cluster, and they may be scattered across a number of buildings within walking distance from each other.' The ideal is that each entrepreneur is actively involved with a nearby college or university business school and is able to participate in team-based business development programmes. Thus, problems of business survival and growth can be tackled in a community of peers. With the assistance of others and with the use of resources of dedicated learning institutions in the immediate locality, the likelihood of survival and of prosperity for new businesses, research findings indicate, is hugely enhanced.

Owner managers in a televillage will have continuous pressure on them, and motivation from their networks, to continue to improve their business and professional skills. When they have mature business development plans, banks and other funding agencies, working with local municipalities and support systems, will be accessible in order to consider whether external finance is required to grow each enterprise. Like the classic business incubator systems, successful televillagers may be able to 'graduate' from their start-up setting into larger premises and higher levels of employees.

Alternatively, many successful micro-enterprises may choose to continue as 1–2 person operations. This flexibility provides, particularly in an age when project teams for larger assignments can be assembled in virtual digital communities from several points on the globe, the capability of responding to opportunities arriving in various formats.

A range of innovative funding mechanisms can be established, unifying the principles of home ownership with those of business investment and creating an equity base for each business startup. Insurance schemes in case of business failure, in which right of occupation can be continued to avoid homelessness, are needed.

Strategic benefits

For many reasons, the televillage concept advances the capacity of older cities and rural villages to respond to the need for local employment and enterprise creation. By removing distance from working and living, the likelihood of regular car commuting is minimised. If large numbers of communities were to adopt televillaging as a dominant form of income earning, far reaching environmental benefits are possible.

Social closeness is also seen as an important in developing entrepreneurial attitudes and skills in communities where such traditions may have been attenuated during the major part of the twentieth century. Here, the social network, both in locality and on the InterNet, becomes a major focus of development. In some cases local communities will produce internationally competitive businesses which can progress to international stature. Conducting their business via satellite links and telenetworking, such firms have the possibility to remaining locally based even as their products reach a global marketplace. With such firms, the revival of both neglected inner city areas and remote rural regions becomes more likely.

Summary: the information economy and metropolitan change

Some emerging economies are confronting the information city head-on by generating new regional structures aimed at innovative information industries. At locations in the advanced industrial world, local experiments are driving toward a new definition of metropolitan development. Whether defined as televillages, telecentres, or networked and distributed corporations, metropolitan economies will be altered by new technologies.

Yet as King's Cross indicated in the 1990s, the old play between local interests and trans-national capital come into conflict. New ways of generating local opportunity and sustaining economic growth are increasingly being sought.

7 The Environmental Entrepreneur: Personality and Culture

Reality Field 7
Entrepreneurial innovators carry the imperative of responsibility for environmental change.

An important aspect of the New Covenant, as new business-community linkages are described here, is the way innovation and change within business are handled. The environment of business is both a biophysical and a psycho-social system of relationships. In coming decades, the emergence of a global business culture more responsive to change and innovation is apparent. This psycho-social revolution is as important as the socio-spatial one governing business and the environment. Chapter 7 explores the cultural, psychological and political background to this development.

This psycho-social revolution can be termed the Entrepreneurial Environment, a complex of beliefs, values and assumptions which is undergoing radical transformation during the years of socio-spatial change on the world's corporate frontiers.

As portrayed in the telematics revolution described in Chapter 6, wealth and power are flowing into new enterprises formed from knowledge-intensive research. Whereas the resources-based revolution of the late nineteenth century formed the global basis of planetary change in the twentieth, the twenty-first century will be dominated by electronics, computing and telecommunications innovators, predominantly located in urban environments.

ENTERPRISE CULTURE IN THE TWENTY-FIRST-CENTURY CITY

Ways of organising work and inventing new environments vary among societies. As the twenty-first century begins, methods of inventing new urban living and working spaces are undergoing a paradigmatic shift.

Initially developed in American business centres and technology firms, these paradigms are radiating across the planet and will displace older approaches. New and revised models lead the development of international management cultures and they show how responsible corporate environmentalism combines with high levels of efficiency.

Reflecting the prosperity ethos in American experience since Emerson, in every society successful entrepreneurs fuse personal will, psycho-social esteem, and economic success. It is these individuals who lead successful urban communities world-wide into the new century, a century where training prospective millionaries is conceivable. In creating new working and residential complexes, individual, environmentally aware, real-estate developers and designers will lead successful urban communities in the new century. In America, Europe and Asia Pacific the new 'edge cities', car-based networks of communities and business parks unified by freeways, airports and train networks, are becoming the core of successful economies.

What drives this phenomenal pace of change? Conventionally, theories about entrepreneurial innovation in technological and information industry are linked to engineering sciences and investment banking. There is a widespread belief that quantitative manipulation is the essence of successful business. But such skills are marginal. On the contrary, there is much evidence that business is also driven by aesthetics and psychology, and a desire to create quality for customers.

The world of events needs a background of sharply drawn dreams before potential solutions can be tested. Alternative approaches, using intuitive, psychological and in some senses literary concepts, are needed. Conceptual tools about organisational and psycho-social change transcend the limits of micro-design. As the Japanese proved over the last half-century, in modern societies product design and manufacture are too important to be left to industrialists and engineers. For this reason, it is desirable to explore the impact of cultural and psycho-social factors in the development of corporate leadership which is not only sensitive to the challenge of improving the environment, but also exhibits the creative and innovative spirit essential for its realisation through innovative approaches.

This chapter addresses three fundamental components of entrepreneurship: a consideration of culture and innovation, a projection of different options for structuring and programming organisations to produce urban business innovation, and finally some scenarios about institutional frameworks for establishing methods for engendering innovation and success.

INNOVATION AND ENTERPRISE CULTURE: PROBLEMS, IMAGERIES AND CIRCUMSTANCES

The way we form images of change and improvement matters. Models of personality and culture provide powerful images. Management models drawn from art, myth, and literature, and from conjecture and invention remake the city. Metaphors allow the analyst some distance from technology. Thus, the means to business success in the new century may lie in art and literature, and in myth.

Just as telecottages are becoming characteristic of the rural countryside, so Internet Cafes are forming a new social nexus in the inner city. Across the advanced industrial world, inner cities are plugging into the Internet and becoming InterCities.

Each piece of urban equipment and household furnishings is acquiring its own set of microchips. The rate of microchip growth doubles each year, yet silicon-based companies' profits are growing only at the rate of 35 per cent each year. The price per microchip is reduced every three years by half. Today, the most expensive part of a car is its computer, controlling in one 'brain' all elements of engine and mechanical operation and sensing. Car producers use more computers than do businesses. The urban environment is not becoming virtual. However, it is becoming re-established in that global landscape which one writer once called 'the non-place urban realm'.[1] This realm of interconnected places now includes the whole of the planet. The central city and its urban neighbourhoods are highly accessible nodes in this maximum cobweb of locations.

Prosperity and creativity in the Emersonian perspective

The twentieth century was declared to have been the American century, but from the perspective of early 1999, it seems that the secular world will be conditioned by an American-driven twenty-first century as well. America was the first country to be organised by secular non-religious groups. The creed which drives American environmental enterprise may be summarised:

> In human affairs, like attracts like. Success adheres to success-radiating persons. The human task is to create with optimism in

> spite of the inherent difficulties of material existence. The material world is essentially invented by human imagination. Every element in society has been conceived, designed and then realised through positive acts of will by inspired individuals. Each individual can choose to undertake self-development to ensure that he remains capable of creating.[2]

The modern tradition stems from F. A. von Hayek's seminal book, *The Road to Serfdom* (1944). In response to this philosophical argument from a leading Austrian political economist, a number of influential groups were formed to underpin the liberal democratic opposition to extremists of both communism and fascism, following the Second World War.

In turn, these groups gave rise to the dominant neo-liberalism of late twentieth-century economics, driven, in particular, by Milton Friedman and the cluster of economists at the University of Chicago and at Yale University, including James Buchanan. Von Hayek and his colleague, Karl Popper, led European theory from the London School of Economics and Political Science.

There were many less institutionalised responses. For instance, a more formal philosophy of personal prosperity and success psychology was developed by the novelist Ayn Rand, whose books (*The Fountainhead*, *Prometheus Unbound*) and essays (*The Virtues of Selfishness*), reflected the prominence of personal achievement based upon radical anarchism and individual liberty in American life. The triumph of Thatcherism and Reaganism during the 1980s gave full cultural esteem and political visibility to supply-side economic theories, resulting in what some call the 'Greed Decade'. Alan Greenspan, chairman of the influential Federal Reserve Board in the US during the 1990s, is a disciple of Ayn Rand's philosophy of objectivism and sustains his faith in the market as a reflection of human nature. The assumption of human rationality, therefore, leads to the conclusion that the market will ultimately be correct. With profound faith in pragmatic analysis, Greenspan argues that 'In analytical people self-esteem relies on the analysis and not on the conclusions.'[3] Defeat of the conservative parties, both Tory and Republican, and their replacement by left-of-centre parties in both the United States and Great Britain in the 1990s, did not significantly alter the emphasis placed on neo-liberal economics, although after the Asian and Russian financial crises in 1998 there was increased pressure on the IMF for some regulation of global financial flows.

In America, ideas of personal prosperity, material success and self-development have been propagated for two centuries, mainly outside the mainstream of academic teaching. Today they persist, still outside the universities. Books, tapes and videos concerning self-motivated success are available in every bookshop and drugstore. This influential but unrecognised literature often has titles resembling that of the most famous book of them all, *Think and Grow Rich*, written by Napoleon Hill and published in 1935.

Napoleon Hill, a young journalist at the turn of the century, was invited by Andrew Carnegie to discover the secret of business success by the famous Scottish-born steel multi-millionaire. His book, the first of twenty-five, has sold over thirty million copies. Using the Carnegie system, Hill himself made and lost several fortunes in the newspaper and publishing business and founded various institutes for propagating these business theories.

Some of his more recent emulators include Ben Sweet, and the latter-day success gurus, Earl Nightingale, Dale Carnegie, Norman Vincent Peale, Anthony Robbins, and Tom Hopkins. Most recently, there has developed in applied psychology an approach based on creative visualisation, called Neuro-Linguistic Programming (NLP), to link specifically desire for achievement and knowledge of brain–physiological linkages.[4]

In the Hill approach, it is essential to make oneself accessible to the inherent creative power of the universe. Acts of prayer, meditation and psychological training focused on positive creation of inspiration, and the creative visualisation of success, lie, these writers argue, at the heart of success for such figures as Edison, Ford, Carnegie and Franklin D. Roosevelt. These writings and seminars have inspired many Americans and others over the last 70 years to create new products and companies. It is conceivable that without this model, less innovation would have occurred. Under these influences, many individuals have built considerable enterprises and accumulated large fortunes.

Nowhere is this culture more in evidence than in the rapid development of the American high-technology sector. In 1995, for the first time in the USA, personal computers outsold televisions. Investment markets began to reflect that reality and by 1997 the stock-market valuation of Microsoft, the leading software firm, was greater than the combined worth of the Big Three automobile giants of Detroit, Ford, GM and Chrysler. As the words 'computer nerd' and 'Silicon Valley' entered the global lexicon, the extraordinary success of a company (Microsoft) which began as a kitchen-table consultancy in Seattle, Washington was revealed. By 1999, Bill Gates had created a multi-billion dollar company which was subject to the ultimate accolade of success, an anti-trust suit in the American courts for domination of the industry.

In late 1998, 20 to 25 per cent of Americans were active Internet users. Industry analysts at Inteco predict that by the end of 1999, US users will increase to 36 per cent, while Europeans continue to lag about a year behind. For example, in Britain, by most estimates, about 10 per cent were active Net users in late 1998.[5] Web resources are overwhelmingly American, forcing Europeans to surf across the Atlantic. It is estimated that up to one-half of the traffic at some US shopping sites was generated from Europe.

Driving this phenomenal rate of technological change is the power of investment. In the IT sector, a stronger risk-taking culture, supported by venture capital, spurs US Internet growth. Mike Butcher, editor of the London-based technology industry newsletter, *New Media Age*, comments, 'Here [UK] even people with good ideas don't have the access to capital that they have in the US.' Despite the high cost of telecommunications in the UK, the predicted increase in revenue from on-line advertising could provide a window of opportunity for creative British entrepreneurs. Development capital is often seen as a one-way street, going from holders of capital into ventures. But in fact it is a system of interchange involving many actors, stakeholders and institutions – a world of persons and their actions, as Figure 7.1 illustrates.

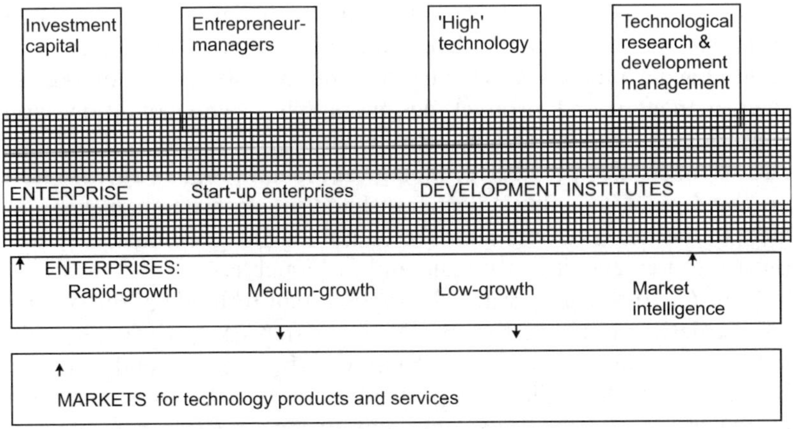

Figure 7.1 Flow pathways: venture capital and technology investment
Source: University of Greenwich Business School, 1998.

The European experience

The development of international venture capital and high technology sectors in the 1990s brought many European initiatives, particularly from the European Union. While Europe is less active on the Internet, it is also true that the Net market is growing faster in Europe than anywhere else in the world. The Net gap is being closed, but the problem of innovative leadership remains. All of the European initiatives in venture capital and high technology have more or less failed to demonstrate major innovation because of their incapacity to link to the psycho-spiritual structures deep within European nations. At the commercial level these structures take their form in marketing and sales management. That is, the capacity of people to spend their incomes on products relates to the value of those products as perceived within individual privileged self-images and lifestyles, the everyday reflection of their spiritual definition of themselves.

But the European models were simplistic. While policy-makers worried about technological superiority, they denied that commercial design and psychological marketing supremacy were the critical other half of technology success. Rather than new inventions, consumer-oriented styling and marketing are required. Better product innovation, rather than more invention, would meet first the American and then later the Japanese challenge. The dominant culture in European countries conditions enterprise formation and investment decisions. European industrial cultures, in contrast with their American and Japanese counterparts, have reflected the belief that the work of scientists and technologists determine the relative competitiveness of technological industries. The truth is far removed: it is marketers, sales managers and product designers/stylists who condition large consumer markets to acquire new technological products. Engineers and scientists may provide technological breakthroughs but these have no commercial meaning unless translated into consumer sales strategies. In the end, not only must the technology be good but consumers must also perceive it as valuable and fashionable.

In order to sustain a viable European high-technology sector, specific initiatives tuned to the special circumstances of particular regional cultures and systems are required. After the expenditure of 2 billion ECUs ($4 billion) from 1975 to 1995, EU investment programmes have produced no apparent European success in any area of technology industry. Only the Airbus and Ariadne space rockets, along with Rolls-Royce aero-engines, GEC flight visual display systems, among global technology sectors, are able to compete internationally. These have succeeded by marketing technological excellence to expert industrial and governmental defence buyers, rather than to mass consumer markets.

European venture capital

The European and particularly the British venture-capital market advanced in the 1990s. Britain could already draw upon the talent of many experienced American venture capitalists resident in London. Long-standing links existed between British and American banks and investment houses.

Venture capital is distinguished from other forms of investment by a number of characteristics:

1. The investment house takes a large, but usually a minority, share of the stock in the company.
2. The house has a hands-on view of how the business should be developed.
3. The house is focused on an exit route from its investment in the period 3–6 years after initial investment.
4. It aims for a portfolio increase of 40 per cent per annum across all its investments.
5. It expects that many of its investments will fail, and a few will compensate with spectacular success, while a few others will survive without prospering.

Investment banks in Europe are used to economies dominated by large companies with market oligopolies in established sectors. Consequently there is little grassroots experience in identifying potential entrepreneurs who could replicate the enormous growth that has been, for instance, a key feature of American micro-electronics.

British bankers and technologists had, for many years, gone back and forth to the States for experience and training. They were aware of the success of the American system. Thus, when circumstances in the macro-environment seemed propitious, the import of high-risk paradigms also became attractive. However, specific features of European society and market economy made such transfers slow to take hold. For one thing, European business communities had in recent decades devalued applied technology in favour of, on the one hand, pure science, and on the other, arts and social sciences. Engineers featured as heroic figures in the Industrial Revolution until about

1880; but thereafter technologists often had to go abroad to find opportunities for achievement.

European countries remain in some ways medieval societies, some with established Churches, monarchies with real if subtle powers, and autocratic forms of Parliamentary democracy. Elitism and class systems permeate key institutions. For this reason, those persons who came from non-middle-class communities and speak with non-middle-class accents suffer the kind of prejudice which in America used to be reserved for newly-arrived minority ethnic groups. Business and politics are dominated by larger, well-established monopolies, corporate entities, which span the civil service, manufacturers, and trade unions. Identified with marginalised and under-educated groups, it has been a tradition that small business is less favoured as a career for able young people.

All this may be changing. As Richard Branson, the innovative and successful entrepreneur of Virgin Airways, noted in a recent interview:

> I do think that there has been a positive change of feeling since this government [Blair] came to power. I assume Blair is talking about creating a classless society, and the country does appear to be moving in that direction. I don't think people consider themselves within their class bracket anymore, but see themselves as human beings. If you are a good leader, then you will be out there drinking in the pub rather than lording it over others. The days of different levels of society staying separate are over, and there is a sense that the middle class is broadening out. Ideally, we need to get to a place where we don't talk about class, like the United States which is a truly classless society. If everyone ends up middle class in this country, then that is largely the same thing.[6]

Most critically, there is a growing pool of under-employed scientists and engineers with entrepreneurial zeal. There are more certain rewards in the professions or in retail and/or service industries. For all these reasons, European investment institutions are losing their scepticism about the possibility of small technological business success.

A technology-park manager in London tells of a Business Plan in the field of microelectronics. It was first put to a City investment bank with names and addresses of entrepreneurs based in Berkshire, near London. The plan was rejected. Resubmitting the same plan, only names and addresses were changed, to those of Dallas-based firms. It received an immediate positive response.

The power of stereotypes, fashionable and unfashionable, in technology fields is as powerful as in any other area of culture. Whether the European investment community is actively seeking to assist cultural change in European industries or whether it is passively seeking to maximise its own position is an issue. To do the second requires a commitment to the first.

ORGANISING AND PROGRAMMING ORGANISATIONS FOR BUSINESS INNOVATION

Models of value differentiation

The values by which investments are made rests within cultures, that is, systems of widely shared values and customs. Among many approaches to building models of investment and technology culture, two general models may be distinguished:

- Model 1: The macro-environment of business investment, affecting the behaviour of capital.
- Model 2: The macro-environment of business formation and enterprise development, affecting the behaviour of individual business managers and their ventures.

The issue for policy-makers, in public and private agencies, is to determine the appropriate mode of intervention into enterprise formation and subsequent success, within both macro and micro-environments. Tables 7.1 and 7.2 indicate some of the macro-dimensions between the UK, American and European experiences in recent decades.

The Mann matrix and European and American modal personality in the technology industry

A further development in the attempt to examine the nexus between personality types, culture and business performance draws upon the insights of psycho-analytical counselling models.[7] Using the typologies of Jung, the psychologist Dr. Harriet Mann developed in *The Pooh*

Table 7.1 The macro-environment of business investment

	Europe 1977	*Europe 1997*	*USA since 1977*
Start-up potential in knowledge-intensive business	Low	Medium	High
Investment capital sourcing	Restricted	Expanding	Broad-based, multi-sectoral
Dominant professional ethos	Accountancy	Marketing	Engineering; law, marketing
Venture capital sources	Few	Many	Many
Securing capital	Real-estate based	Business plan + real estate	Business plan + real estate
Higher education linkages	None	Multiplying	Pervasive
Power of banks	Monopolistic	Reducing	Market-structured
Support of taxation system	No	Yes	Yes

Stakehold, a provocative, but not necessarily operational, concept for comparing different modal personalities relating to industrial culture. Mann allows us to identify a range of viable personality styles operating in management settings. She suggests that up to fourteen (sixteen minus one, as two of the possible combinations are psychologically inconceivable) modal personality types are operating at any given time.

Using the Jungian typology, the well-known Myers–Briggs inventory developed a method of psychometrics for testing individuals. The Myers–Briggs approach establishes, through written questionnaires, a ranking of individual preferences and responses, along four basic opposites: (1) extroversion/introversion; (2) sensing/intuition; (3) thinking/feeling; (4) judging/perceiving.[8]

The Myers–Briggs approach results in a very complex matrix of 16 personality types which are based on the six variables in Table 7.3 and which are ranked in importance depending on the answers provided in the individual's questionnaire. Characteristics associated with each type are then identified and related to team performance. The Myers–Briggs system is widely used by career counsellors, employment services, and consultants who conduct training programmes for corporate clients, especially those who are working in a cross-cultural environment. One such consulting group in London, formed by two American women and managed by an international team of eleven women and one man, has delivered customised training programmes for companies such as

Table 7.2 Macro-environment of enterprise formation: psycho-cultural mode differentiation*

	UK 1980	*UK 1997*	*USA since 1975*
Psycho-political value system	Low growth; stable career; large employer systems; small business discouraged	Medium growth; career changing; rapid moves among large employers	Fast growth; multiple careers; small and large employers; new enterprises revered; the privileged paradigm
Expression of emotions	Prohibitions against display of emotion	Organisations increasingly reward display of emotions	Heavily rewarded in organisational culture; strong motivational system built around high-affect displays
Organisational paradigm	Hierarchical, bureaucratic	De-layered; secretive	Decentralised, atomistic
Emblematic organisational paradigm	Balance sheet	Corporate strategy	Business plan
Relationship modality	Vertical loyalty	Professional versatility	Holistic and mutualistic involvement
Central personality competence	Autocrat	Technical guru	Creative artist; technological whiz kid

*In this model, small-group and internal psychic phenomena determine the viability of different business propositions and can be distinguished.

Table 7.3 The Myers–Briggs type

Extroversion	*Introversion*
Sensing	Intuition
Thinking	Feeling
Judging	Perceiving

Andersen Consulting, Unilever, Motorola, KPMG, Price Waterhouse, Morgan Stanley, the British Standards Institution and Philip Morris. As our case study demonstrates, analysis of the entrepreneurial spirit and business success has often been cast in terms of a combination of leadership qualities and cultural influence.

The Silicon Valley case

Silicon Valley was in the 1990s a network of entrepreneurs, technologists and investors, interacting together with a high volatile manner to create some 1,500 successful microelectronics firms and unnumbered failed firms.[9] According to Larsen and Rogers, it has also been heavily dependent on non-unionised exploited female labour, non-interference from environmental health officers, and few planning controls. Environmentally it is congested, racked with ground, water and air pollution, and socially a polarised network of haves and have-nots. However, so powerful is its economic success that few governments would be sorry to have such problems in the suburbs of one of their major cities.

For managers, the pressures of time commitment, stress, mobility, anticipatory time, and fast-footedness are intense. Divorce is as common as marriage and weekend working is the norm. In an urban geography of freeways, coffee shops, and low-rise factories, the power of geography is reminiscent of the City of London and Wall Street. Face-to-face relationships are the most powerful form of communication. Jobs and careers take new and unexpected pathways: who meets whom over breakfast at a local cafe, who drinks beer with whom after work and at the weekend beer-and-barbecue party.

The key personalities in this venture ecology are liable to be gregarious, risk-seeking, interested in but not obviously greedy for personal wealth, and unwilling to settle for small victories. In the Mann matrix, many are explosive ('volcanic'), emotionally rich, extroverts. One venture capitalist says, 'Give me all the raving egomaniacs with IQ's of 140' (quoted by Larsen and Rogers). The Valley syndrome favours technological mavericks and egocentric behaviour. California's Silicon Valley, for instance, gives precedence to teams which feature three of these types:

- the feeling volcanic: the marketer
- the intuitive volcanic: the venture capitalist
- the thinking territorial: the scientist

In contrast, The City of London can be said to feature:

- the sensate ethereal: Oxbridge classicist/merchant banker
- the thinking territorial: financial dealer/empire builder

Yet the key technological breakthroughs – microchips, integrated circuitry, ingenious software – have often been developed by reclusive introverts with very high IQs – the 'nerds' with whom Bill Gates, the most famous entrepreneur of them all, identifies. Technology firms may be started by brilliant engineers, but only rarely can they grow to substantial competitive size without a team concept, as the British psychologist Meredith Belbin has described.[10] Brilliant designers are rarely good marketers, personnel managers or production managers.

The unique contribution of Silicon Valley as a specialist sub-culture is thus seen to be the ability to form spontaneously large numbers of teams combining imaginative capitalists, brilliant inventors, and human relations managers, all focused within a single industrial sector. Not only are numbers of teams readily formed, but they are formed repeatedly in new combinations when old teams do not work. Just as investment follows success like a flash, it also immediately abandons failure.

The overall success of a few winners is so great that the investment community can tolerate failures. Yet in 1982, after the worst recession in America for fifty years, the five-year survival rate of Silicon Valley ventures was over 75 per cent. This figure contrasts with small business survival rates in the USA, seldom over one-third after five years.

Within the USA, there are many institutional burnt fingers over the last fifteen years. Investment in urban real estate, agricultural property, and home loans have at different points come unstuck. The high-tech venture capital market will at some point have its downturn. All markets, without exception, move in cycles. For this reason venture capital was too easily available in the USA, funding many imperfect business plans. The recession of the 1980s in microelectronics showed the weakness of many such plans.

By 1999, however, the US economy was booming. With the lowest unemployment rate in 40 years (4.3 per cent), low interest rates, food and fuel prices at an all-time low price, low inflation, average earnings rising (3.8 per cent in 1997), and a budget surplus of $700 billion over the next ten years, US growth continued to advance, as did the strength of the IT sector. Combined with dramatically dropping crime rates (lowest murder rate in three decades), teenage births declining, the lowest infant mortality rate ever, a stabilising divorce rate and life expectancy continuing to rise, the United States demonstrated once again its extraordinary resilience and responsiveness to the need for change.[11]

The American stress on risk, individualism, and creative competition among firms and regions makes central planning and forward control of innovation disreputable. Only in the late 1980s has even the mighty electronics industry turned to Federal protection to offset the 'unfair advantage' of Japanese industrial co-ordination. This follows previous American government interventions in aerospace, agriculture, petroleum, health care and car building, where, in spite of anti-government sentiment, large businesses were able to marshal government support, particularly in fair trade issues.

The West European and Japanese approach

Technology enterprise within large bureaucracy, by contrast, is at the heart of the non-USA policy systems. Government and private industry are involved in methodical pursuit of development stretching over decades. The planning ethos is non-spontaneous and despises co-ordinated entrepreneurialism. Top-down success in technology can be identified and funded, as in the Concorde project (now successful in economic terms), the Airbus (becoming successful), and the Ariadne space vehicle launch programme. Europe's top-down managerial style, particularly in the United Kingdom, has its origins in historical social stratification, an elitist educational system and the profound distrust between the aristocracy and politicians on the one hand, and industry on the other. Martin Wiener's seminal book on the decline of the industrial spirit in Britain between 1850 and 1980 sheds important light on the ethos of the public school which disdained science, and of the universities, where a career in industry was discouraged.[12] While this top-down approach has achieved some success, it is also a barrier to the sort of informal, free-form creativity which was so characteristic of American technological advance. The dangers of over-concentration of authority and control in Japanese top management has been amply demonstrated by the banking collapse in Japan in the 1990s.

Comparison of American and European team management

Successful businesses require more than one person, for there are at least eight work areas each needing different personal skills.[13] When several managers are working together towards a common goal, the notion of 'team' becomes applicable. Team management is the process of gaining the benefit of more than one person on a particular task. Working in teams requires a method of communication and interaction which is informal; that is, teamwork requires both emotional and intellectual open communications.

Informality at work in America is general; and therefore teamwork is easier to establish. In some sense or another, America was founded on a common rejection of the formalism which characterised European societies. Formality in Europe gives structure and meaning – the social hierarchies, the subtle psychological distinctions among classes, the pre-eminence of 'high' cosmopolitan culture over folk and vernacular cultures.

In business and industrial life, these formalities can discourage international companies' capacity to adapt to the rapid-change climate of high technology and product innovation. The kind of personality which can interact creatively and produce good business ideas in team settings does not respond well to formal agendas and linear bureaucratic procedures. During the 1980s, there were moves toward less formal and more spontaneous and open communications both in government and in business.

Many American companies, of course, are bureaucratic, not least of all IBM. Indeed, as IBM became more formalised, it lost its capacity to dominate markets. Companies where formality has taken over have suffered severe loss of market share: Witness the dismal profit records of Ford and General Motors during the 1980s. New products must be developed in informal settings: This fact was crystallised in the early 1980s when IBM, recognising the problem of their internal culture, set up an insulated product group, a 'skunk works' in Silicon Valley jargon, to design the IBM microcomputer. This group worked in the laid-back style of Apple Inc. and Intel and the other Silicon Valley companies and were able to come up with a product which, while not technologically advanced, could be marketed by IBM.

Sales management – the forgotten component of European business

Large successful companies in a quasi-monopoly position in their industry often discourage entrepreneurs. Technological supremacy was not the essence of IBM's success in the micro field, nor indeed was Apple's. These companies have continued to innovate successfully by a combination of marketing, pricing and product innovations around machines which did not compare well with some of the better European products. What has worked in these companies' success strategies is advertising, marketing and sales management. The key to good sales management is not invasion of customer control or privacy. Instead, it is detecting those parts of the potential customer base which might buy if presented with a desirable product at the right price; and then reaching them with a decision point which is attractive to their pocket-book and need.

The focus on manipulation and near-coercion in high-pressure sales is rightfully anathema to the effective sales organisation. It is a distortion to view sales management as an area of moral compromise. Rather, industrialists and technologists, who dominate industrial development research in some European settings, must develop a commitment to customer response, if the Japanese capacity to study customer needs and respond rapidly to them is to be confronted.

Team management is focused on that crucial point of sales decision by the buying public. Enormous investments are made in discovering not only what technology can provide but what customers want to buy. The technological pre-eminence of American and Japanese enterprises comes not from a fixation on having the best new technology but in focusing marketing and service organisations at an early stage to changing customer potentials and desires.

Creativity in management and in science often stems from the energies of more junior and younger members of staff. The teams which can best use their energies will be lateral and informal, will use brain-storming, will permit display of emotion, including anger and joy, and will reward creative activity without heavy overloads of critical experience.

Explanations for the failure to focus on marketing and service have centred on the lack of adequate numbers, quality and remuneration for scientists and engineers. The key variable in advancing technology sectors, however, lies in promoting a more responsive type of business enterprise, whether public or private. Responsive business values allow technologists and managers to interact in creating new strategies in a more fluid and positive way.

In spite of these pressures, there is a large pool of under-employed engineers with entrepreneurial zeal, but without the communication skills, college networks, class-linked behaviour and social confidence to develop sound technology businesses. Low wage levels, allowing the use of low-paid labour in electronics assembly, suggests that some of the reasons why the Third World attracts investment also apply in North West Europe.

The summary in Table 7.4 of ideal company types, personality traits and management cultures presents a method in which systematic comparisons can be made among companies at various points in time: when they are on the rise; at the peak of success; and in decline.

Table 7.4 Company types and management culture

	Sunrise	*High noon*	*Sunset*
Leadership style	Balanced between creative and implementor types; meta-rational and entrepreneurial	Dominated by implementors; technical-rational, consensus seeking	Dominated by completers; seeks fail-safe solutions
Strategy	Decentralised	Centralised	Fragmented
Senior management motivation	Achievement	Affiliation	Affiliation
Team interaction character	Informal and egalitarian	Formal and procedural	Bureaucratic
Mode of handling internal innovation	Enabling	Monopolising	Suppressive
Shop-floor management style	Developmental	Confrontational	Rule-driven
Quality management systems	Continuous	Improvisation	Decadent
Strategic portfolio management	Seeks 'stars'	Revolves around 'cash cows'	Full of problems and 'dogs'
Marketing culture	Customer focused	Technology focused	Producer focused

SCENARIOS FOR INSTITUTIONAL FRAMEWORKS TO GENERATE PREDICTABLE INNOVATION AND SUCCESS

The Japanese use a system of low labour costs allied with deliberate encouragement of ferocious competition among large firms. This form of technological development can be seen to be workable in European settings, best evidenced by technology firms. Emphasis on government intervention, on deliberate corporate innovation, and on large-business domination of the economy indicates the need for government policy interventions:

- a central technology policy and industrial investment ministry;
- a central industrial banking agency, with preferential rates and special funds for technological innovation;
- a premium awarded to able students undertaking technological training, and to institutions able to train students in the disciplines of new enterprise development.

Innovative enterprises have traditionally emerged from within an existing structure of firms and institutions. Two examples are apparent: The City of London's long-term adeptness in devising new money instruments is a signal area of product innovation over centuries. German and British chemical giants, all founded in the eighteenth and nineteenth centuries, became world leaders in their industries.

Technology culture and national characteristics

High-affect behaviour characteristic of Silicon Valley culture enables ordinarily introverted personalities among creative technologists to communicate new ideas without the rigidity of bureaucracy to hamper their development.

Present successes have been in second-generation technologies requiring long lead times and perhaps a lower level of sheer creativity than in microelectronics. The thinking-territorial technologist comes into their own in this more controlled and planned entrepreneurial environment.

Therefore, future successes may well be those in which, like Japan, deliberate buying-in of major technological developments and their commercial development will produce better results than, for instance, cultivating highly creative first-generation science and engineering breakthroughs.

A major re-focusing of government expenditure away from 'pure' science and into applied business development using new technology is needed. The great cost of major nuclear physics experiments seems no longer to promise any commercial lead to European countries. Instead, there must be a major emphasis on import substitution, adapting the typical market preferences found among European consumers to ensure that European technology enterprise can capture a higher proportion of their own markets than is now presently the case.

This will mean spending money on business management schools, rather than scientific establishments. Research on technology enterprise/development, rather than technological science, is needed as well as a premium awarded to able students undertaking technological training and to institutions able to train students.

A central technology policy and industrial investment ministry combined with a central industrial banking agency could be created with preferential rates and special funds for technological innovation.

The scenario of Environmental Development and Innovation Centres (Box 7.1) summarises the primary features of a proposal which would unify community, work and business skills. It was developed in response to the demands of globalisation which are expected to increase in the twenty-first century. It was also inspired by turn-of-the century proposals for integrated living schemes as well as the challenge of regenerating dilapidated and abandoned industrial areas, particularly those along rivers and in ports. The Docklands development along the Thames in London has realised parts of this scenario, as has the new marine and housing development in Brighton.

Box 7.1 Scenario: Realisation of a chain of 'Environmental Development and Innovation Centres' (EDIC)

The planet will unify in a multi-national and multi-cultural system the energies that have been available only in a limited way during the majority of the twentieth century. A particular multi-use, community-based approach will be found in the development of residential and office communities, focusing on special waterside clusters of offices, shops, schools and housing.

The global dimension

Each community will have firms representing several countries of Europe and operate in several languages. There will be a premium on multi-cultural marketing and service development. Each of the businesses in EDIC will be a service-directed international marketing group. Most will operate as small-scale enterprises; some will be linked through joint marketing arrangements to be franchises of larger groups. The common factor will be the emphasis on services.

Integration of retail and small-scale production workshops

The EDICs will have a range of craft workshops, small technology production units, and offices, ateliers, studios and other information-processing workshops.

High-growth design-led small enterprises

In many respects the EDIC will be an incubator for firms which can grow rapidly. These firms will grow through replication of other small enterprises, using franchise techniques and small multiples. Thus, entrepreneurs will be able to retain the benefits of the EDIC without sacrificing the attractions of growth and economies of scale.

The recycling of obsolete industrial and commercial properties

The least expensive place for these EDICS will often be in obsolete small industrial towns near coastal locations. There will be an associated emphasis on product and service design and the need for using high-amenity environments to market and project this business.

Programme specification:

1. *Capital requirement*: $60 million/£40 million
2. *Objective*: To develop at least 10 properties as multi-use business sites combining:

   – special professional and health care facilities
   – design offices
   – art and architectural studios
   – gallery
   – specialist and craft shops
   – business conference and training centre
   – residential/hotel
   – bar restaurant
   – mooring and landing
   – development of flats and apartments

3. *Business structure*: Business will be as a limited company with several investors, backed by bank facilities, with the freehold to be disposed to institutional investors at completion.

Training the next generation of industrial millionaires

The essential opportunity facing global business communities over the next two decades is in finding a method of producing the industrial fecundity which has been dominated first by the USA and more recently by Japan. The methods of doing so have been studied and applied by Japanese technology industries and linked government departments. In the new global setting, the system in which new businesses start up consists of:

- the business community
- the banking and investment community
- the higher and further education institutions
- voluntary charitable institutions and foundations
- government ministries

The appropriate method of working forward toward a generation of millionaires educated in technology is through the designation of specific Enterprise Institutes serviced jointly from these various institutional sources. Box 7.2 suggests one such approach.

The enterprise development process: summary

The procedure used for developing millionaire-producing enterprises can be formulated as in the proposed diploma course (Box 7.2). Its key features include the simultaneous development of increasingly more refined business plans and procedures in each of four areas: Management, Marketing and Distribution, Technology and Finance. Using a sequence of rapid repetition of plans, developed within a nucleus management team, a new business can be established capable of rapid growth. Its key features will be flexibility and comprehensiveness of planning and management systems. In this respect it will use state of the art business modelling software.

DEVELOPING A GLOBALLY COMPETITIVE NEW BUSINESS IN A WORLD OF CORPORATE ENVIRONMENTALISATION: 12 PRINCIPLES

There are 12 principles which if followed consistently will assist in creating a successful business that functions at a high level of environmenta-

Box 7.2 Profile for an Enterprise Institute diploma course

GLOBAL ENVIRONMENTAL INSTITUTE FOR BUSINESS DEVELOPMENT AND SUCCESS PLANNING

The Diploma Course in Millionaire Creation

Course Objectives:

- To provide participants with the understanding and practical tools required to create a personal strategy for establishing substantial wealth
- To provide a group setting in which participants work with one another towards their respective personal strategies

Location: World-wide
Access: Open to the public, including part-time students
Scheduling: One year part-time, including two weekend workshops; three evening per month for ten months, one evening per month for two subsequent years.

lisation. These twelve principles are summarised and discussed below:

1. Dematerialisation
2. Distributed networks
3. Telecomputerisation
4. High knowledge content
5. Micro-structure
6. One-person bands
7. Continuing professional development
8. Continuous innovation
9. First time right every time
10. Delighting the customer
11. Cherishing business partners and stakeholders
12. Pervasive marketing: the three rules of entrepreneurial success.

1. Dematerialisation

Concentrate on products which have little material content and enormous intellectual content. Even if you are not expert in a particular field, you can become a 'publisher' for others who want to specialise.

2. Distributed networks

Create a personal network of colleagues, friends and co-entrepreneurs. Use electronic and snail mail to keep them in touch with your bright idcas.

3. Telecomputerisation

Get wired up. Your key clients are and you want them to be able to e-mail you. Furthermore, you want to be in the position, as the Internet develops, to generate specific websites with your favourite customers. Consider when you can afford to establish video conference links with your clients.

4. High knowledge content leads to being a high-end operator

Work in products where your professional training and development count for a lot. Don't get trapped in specific expert areas which for a while command a high premium and later on become 'industrialised' so that less senior people can deliver good quality at a much lower price. Being a high-end operator – a highly paid international expert – means moving your field rapidly forward.

5. Micro-structure

Become expert in specific how-to considerations in your special area. Establish a reputation in a particular technical area and get known for being able to handle difficult bits in a larger project.

6. Be a one-man band with dozens of partners

Do not grow a company of employees. Instead, grow a company of high turnover with many sub-contractors. Develop a stable of slightly younger people, or older semi-retired colleagues who you work well with, and with whom you have large areas of professional agreement. Then, writing proposals and pitching for work will become a matter of putting together the 'old' team and having fun.

Where you need technical backup – secretarial, communications, co-ordination, bookkeeping – find a local firm which offers packages appropriate to your needs.

7. Continuing professional development

Never stop learning new skills and enhanced capacities. The calculation of the half-life of any professional training today is between 3 and 6 years. In other words, within about 5 years half of what you have learned in a comprehensive professional training will be seen as obsolete by the profession. You cannot afford to be out of touch with the changing face and soul of your professional community. So you must see training as a continuous activity, involving books, TV, courses, conferences and sometimes a further named qualification at college.

8. Continuous innovation

Work continuously on improving your service and product. This is a given in today's competitive world.

9. First time right every time

Deliver high-quality products on time. This is the famous law of Edward Deming, which in some ways has brought about a global revolution in quality product and services, led by the Japanese and now permeating work in every country where people aspire to high standards of living.

10. Delighting the customer

Deliver more than the customer expected and go beyond the contract, when you know the customer will be receptive to the experience. Steve Perham, a colleague and co-writer, often asks as he enters a meeting: 'What can you give me which will cost you absolutely nothing and which will mean the world to me? What can I give you which costs me nothing and which will mean the world to you?'

11. Cherishing business partners and stakeholders

Value your sub-contractors even more than you do your customers, for they are your family. Richard Branson, one of the most successful of global entrepreneurs today, says it all: 'I look after the people who work in my organisations and they look after my customers. By paying attention to these people, I don't have to worry about my customers: they will be delighted.'

12. Pervasive marketing: the three rules of entrepreneurial success

In personal career development, the clear edict is that one must choose work which provides maximum satisfaction and full human relationships. That is, to choose work for reasons which are financial or security is almost always destructive of highest creativity and personal potential. The reason is simple: in order to achieve high rewards, one must be fully motivated and emotionally committed. To feel this commitment, one must work out of team loyalty and inner devotion. Consider the following simplistic equation:

Achieving quality work	Do good.
Achieving quality recognition	Be seen as good.
Achieving quality compensation	Do well.

The rule is never to let professional and commercial networks forget what you are doing. Spend as much time on marketing as on crafting your product. Expect the best rate of pay in your industry for your contribution, consistent with the top going rate.

Achieving high-quality results takes about a third of working time, leaving two-thirds of work hours for two further activities – gaining recognition for what has been achieved; and getting paid at the highest appropriate rate for the work. Top people in every sector are the ones who have understood this 'Iron Law' – 80 per cent of rewards go to 20 per cent of the professionals in any business.

To succeed at the top level, it is not enough to be good. You must be recognised as among the best and you must be properly compensated within the norms of your profession. This means being in touch with all those people who can advance the interests of your work. And this means taking time and trouble to go and see them, on their home ground. It means knowing your subject. More than that, you must get to know the influence leaders who form the consensus about what is good and who ends up where.

Gaining global recognition

Three rules for gaining recognition as a leader with global potential in your field include:

1. Communicate through multiple channels, including one-to-one meetings with opinion-formers, press releases, books, articles, speeches, conference papers, joint research presentations, audio tapes, videotapes.
2. Write for the international, national and local press. The papers get read.
3. Be active in your professional and industrial associations and institutes. Become a leader by taking on small tasks.

Strategy for career building

Organise the corporate work programme as though you were a one-person business, with your customers inside as well as outside the organisation. For example, the following method could be utilised:

- Send out a newsletter from time to time about the work you are doing.
- Ensure it reaches a wide array of managers inside the company.
- If you are independent, use it as a means of reminding your customers and contacts that your work is helping advance the field.
- Regard yourself as a communicator

Gaining world-class compensation

Four rules for getting the right level of compensation include:

1. Establish your daily rate of pay at the same level as the top 25 per cent of the profession in your field. You can discount this back for special customers; but start with the headline rate. If you have come into an organisation in a lower-paid role, let them know in a professional manner that you hope to progress up the pay scale.
2. Keep up a rapid rate of applications for new work and/or new roles. Show you are mobile and attractive to other customers.
3. Present yourself and your work in comprehensive up-market packages. Dress well and use the best materials and methods for your products and services. In many ways, you are the product of your dress, and grooming should reflect the highest quality. This does not restrict you from adopting unconventional solutions where the culture of your customers rewards stylistic innovators.
4. Show that you are of an international standard in your work by carrying out all the methods of communication and publicity suggested above.

THE FUTURE OF THE ENTREPRENEUR

Big business will be around for many centuries to come. But early in the twenty-first century the great opportunities are in forming smaller businesses and innovating new products at a global level. This can be done in environmentally sensitive ways, in every industry. In general we are replacing the task of moving people and goods with the task of moving information.

Smart chips are being built into every form of technology and service. All these will require a wide range of new businesses which are generating ideas and innovations. Linked into professional networks and informal webs of idea interchange, the coming generation of business innovators will simultaneously be working on environmentally high performance systems using low-environmental impact methods.

The dangers of environmental deterioration are well documented. The opportunities for social and techno-industrial innovation to reverse this trend are even more impressive.

8 Corporate Innovation in the Fast Lane: Environmental Technologies and Planning the New Industrial Metabolism

Reality Field 8
For large industries in coming decades, including cars and the building sector, environmental technologies provide a major source of profit and strategic improvement.

Some of the largest new industries, like software and telecomputing, have profound implications for the environment. They will in many sectors make movement of people and goods between locations unnecessary, thereby reducing the largest source of transport pollution. Moreover, increasingly we will miniaturise and computerise all of our material technology. In an attempt to give increased salience to the impact of developing environmental technologies to corporate performance, we have chosen three case studies for investigation and analysis: the environmental car, the environmental house and the environmental factory. In all three cases, rapidly developing technologies have been given greater impetus by the promise of enhanced competitiveness as well as by a growing awareness of the finite limits of the old industrial and social approaches.

CASE: THE ENVIRONMENTAL CAR

The environmentalisation of the car as the largest single technology industry on the planet has been going on since the oil crises of the 1970s. By the 1990s its fuel efficiency progressed some 33 per cent overall, with some smaller models 200 per cent more efficient than comparable models of twenty years earlier.[1] Engine improvements included multi-valves, fuel injection and electronic management systems. In 1995 American car manufacturers built more computers into cars than were built for the whole of the rest of the non-defence market.

However, car buyers moved progressively into higher levels of convenience, comfort and safety, requiring extra weight of equipment for such items as air conditioning, safety features, catalytic converters, crash airbags, and other elements. With every advance in engine efficiency has come new consumer preferences and government regulations requiring extra on-board equipment. In terms of fuel consumption, then, the mainstream car for advanced economy markets was projected to be by 2010 not radically different. Although the number of workers in the automobile industry has declined, the production of cars has risen in Europe for the past decade. For example, in 1982 Britain produced 877,000 cars and in 1998 that production rose to 1.7 million, despite a persistent overcapacity.[2] Similar figures can be cited for Europe as a whole. Research by the Automobile Association suggests that the growth in car ownership has been caused by combination of factors: increased wealth, two-car families, older people driving later in retirement, more young people passing their tests and staying single longer. It also suggests that this growth may have peaked.

On the positive side, the main construction material for cars will permit a range of weight improvements. Both advanced high-tension steel and aluminium bodies may lighten the car to some 60 per cent of its present weight. The use of complex polymer plastics as high-quality structural elements will also lower weight.

Even more promising is the potential for non-petroleum electric cars in cities, where the main sources of pollution from the internal combustion engine occur.

The prospect for electric cars

Electric cars are not a new idea, but they may well be an idea which is finally coming into its own. Although Joseph Henry invented the first d.c.-powered motor in 1830, Thomas Davenport is credited with building the first practical electrical vehicle (EV) in 1834. Development quickened thereafter with the first two-passenger electric car in 1847 and a 20-m.p.h. car in 1851. Gaston Plante's 'rechargeable' battery in 1859 resulted in an increased range, and in 1899 the first land speed record of 66 m.p.h. was set. One year later in 1900 an electric car was driven 180 miles on a single charge. Charles Kettering's electric starter, invented in 1912, began a decline, however, which was fostered by Ford's mass production of gasoline-powered cars and an abundant supply of inexpensive fuel.

By the 1960s, the negative effects of the gasoline engine were becoming increasingly apparent as concern for air pollution (1960s) and supply (oil crisis, 1970s) mounted. EV development is driven by the market for fuels; if fuel is abundant and cheap, interest in electric vehicles tends to wane, as was the case in the 1980s. However, the technology of car batteries and the ratio of power to weight reached commercial viability only in the 1990s. For the car business worldwide, the electric car promises that something approaching environmental sustainability for the industry can be achieved.[3]

For and against electric cars

Today in the United States, Ford, Chrysler and GM have advanced electric car programmes and some independent EV companies, such as Sebring/Vanguard, were attracted to the potential market. Sebring/Vanguard produced 2,000 CitiCars and was at one time the fifth largest automobile manufacturer in the US.[4] Small companies such as Solectria and AC Propulsion continue to produce electric cars and buses, assisted by subsidies from the state of California to encourage their development. The city of Santa Barbara in California, for example, has one of the largest fleets of electric buses with fourteen in service.

Although the purchase price of electric vehicles is higher than cars powered by the internal combustion engine, they have lower maintenance costs, are less polluting, quieter and are able to accelerate very rapidly. On the negative side, however, their range is limited (50–130 miles before their batteries need to be changed), and while they reduce air pollution, they may increase the use of coal and nuclear power to produce electricity for the batteries, although the latter problem is solved by using fuel cells powered by solar-powered hydrogen and also by cleaner power plants. For example, coal plant emissions are now heavily monitored and under stringent EPA restrictions. Conversion of existing cars to electricity is also much less expensive than new electric cars.

Critics fear that new power plants would be required if the number of electric cars were to increase substantially, but Southern California

Edison estimates that it currently has enough slack night-time capacity to recharge 600,000 electric vehicles within its service area. It has also been suggested that additional pollution from power plants would be the trade-off, but according to the California Air Resources Board, EVs emit 90 per cent fewer pollutants than gasoline, even after including emission from the power plants. A more serious barrier to extensive EV production plans remains the cost of the vehicles, although Solar Electric says there are now several four-person models available for under $10,000.[5] GM's EV-1 was priced at $33,395 in 1997. It is to be expected that prototype vehicles, not yet in mass production, will be more expensive. Solectria's Sunrise, for example, cost $100,000, but once the car is mass produced the price will be in the $20,000–$25,000 range.

Green cars: an emerging market

By 1999, 300 GM EV-1s and 50 Honda Plus electric cars were being leased in the US, and Edison EV in partnership with utilities in Southern California and Arizona have installed 400 charging units in 200 locations. The GM EV-1 will accelerate from 0 to 60 in just nine seconds and a modified version of the GM Impact broke the land-speed record for EVs in 1994 with a top speed of 183.8 m.p.h. Both the Honda Plus and the EV-1 also have regenerative braking, where the energy lost in braking is recovered and converted to electrical energy to charge the batteries and thereby adds 5 per cent to the range. More recently, a composite-bodied Solectria Sunrise went 373 miles on a single charge using nickel-metal hydride (NiMH) batteries.

In late December 1998, Nissan announced initial delivery and pricing information for its Altra EV, the zero emission vehicle developed for use in the US. It combines the features of a compact van with the environmental benefits of an electric vehicle. The 1998 demonstration fleet consists of 30 vehicles with 98 additional planned for the year 2000. Fifteen of the current fleet will be available to fleet customers and Nissan employees for testing and evaluation and fifteen will be leased to four California utility companies: Southern California Edison Co., Pacific Gas & Electric Co., Los Angeles Department of Water and Power and San Diego Gas & Electric Co.[6] The Altra EV, manufactured at Nissan's Tochigi assembly plant, is the first production EV in the US to use a Litium-ion (Li-ion) battery which was developed jointly by Nissan and Sony Corporation and offers a lighter battery pack, a range of 80–100 miles and takes only five hours to recharge.

Budget Rent-A-Car Corporation and EV Rental Cars (Los Angeles) are now offering the first EV rental fleet in the US. It includes a range of EVs manufactured by DaimlerChrysler, Ford, GM, Honda and Toyota which can be rented for $40.00 per day from the Los Angeles International Airport. Edison International provides an electric-powered float for the Pasadena Rose Parade each year which is a production of the stockholders, not paid for with ratepayer funds. In late 1999 the American Honda Motor Co. will unveil the prototype for a new hybrid vehicle (VV) and a new roadster prototype (S2000). As Tom Elliott, American Honda executive vice president put it, 'The VV reinforces Honda's long history of bringing environmentally responsible vehicles to market ahead of the competition.'[7] Indeed the VV is engineered to meet California's Ultra Low Emission Vehicle standard, currently the world's most stringent one. It is a two-seat personal coupe made of aluminium and plastic body panels, weighs less than 200 pounds and will double the fuel economy of the leading gasoline-powered cars on the market. Honda, however, may not be 'ahead of the competition' by very much as Toyota is already producing a gasoline-electric hybrid car called the Prius.

The major Japanese and American automobile manufacturers are clearly positioning themselves for a future market. In the US it is estimated that 90 per cent of all second car driving is under 21 miles per day and 75 per cent of the daily travel is within 31 miles of all drivers' homes. Many of the 4,000 electric cars currently driven are conversions from existing cars and are used as second or third family cars. Given the increasing congestion and pollution associated with transportation, the incentives offered by electric cars remain attractive, despite the current limitations.

In Europe, sustainable approaches have been adopted to support the French car industry's development of a wide range of electric options available to consumers. For the major producers, Renault and Peugeot-Citroen, the aim has been to produce electric versions of long-established models. In addition, dedicated utility vehicles running on electricity are being produced for major users such as the national power company, Electricité de France (EDF). Typically the new electric models are priced at about the same as internal combustion version models. Their acceleration is markedly slower. However, as their major use is in urban locations, this is no problem. Nor is their approximate 90 miles/160 kilometres distance per charge. From flat the current batteries (1995) require 7 hours to attain full charge.

In the city of La Rochelle on the Atlantic coast where the cars have been in use since 1993, electricity charging stations have been placed on kerbs across the town. Other systems under development include an annually subscribed take-and-leave small urban bubble-car called Tulip (Transport urban libre individuel et public). Remote handsets allow access and they can be picked up and dropped off in larger cities. Subscribers aim the handset at the green light on the roof of the vehicle. If it lights up, the vehicle is available for hire, the Tulip is rented, the doors unlock and the car may be taken away.

The French experiment has been supported by substantial government incentives and direct interventions from major energy suppliers. Technology costs are projected to fall over coming years at a rate of 12–15 per cent per annum. By the turn of the century it is conceivable that private investment will find it attractive to commission fleets of electric self-driven vehicles, as well as car production for ownership.

Car manufacturers may be able to expand their business into developing countries with low incomes and poor environments. The car's undoubted benefits in personal mobility and regional decentralisation can be achieved without some of the costs associated with road pollution. France indicates how rapid progress can be made in this area: by the mid-1990s, some 6,000 or more electric cars were in use in French cities. Government subsidies for electric cars were about 20 per cent or $3,000 US, in the 1994 programme.

CASE: THE ECO-HOUSE AND THE ENVIRONMENTALISED TOWNSCAPE

As in the case of cars, photovoltaic (PV) solar battery technology is rapidly accelerating in efficiency. For the first time it is becoming practicable to plan for wholesale substitution of conventional energy systems in houses by solar-powered systems. In Oxford, architect Susan Roafe has built a conventional suburban home entirely powered by solar electricity.[8] On its garden side, 48 PV panels will generate electricity to cover all energy requirements in the house – heating, lighting, TV, video, dishwasher. Special equipment includes a converter to turn DC power to AC, as well as metering. Maintenance is low. Another interesting feature is that the PV system is positioned on top of the roof and thus can be 'retrofitted', that is, fitted after the house is built.

Capital cost of energy in the mid-1990s was $38,000/£25,000. Against this can be placed a projection of the total costs of energy in a conventional house, which may be in the order of £1,200 per annum. Thus, the payback period is quite long: 21 years. However, the costs may fall with the increasing production of PV systems worldwide in coming years. When payback reaches 5–7 years, it is probable that many thousands of households will find it advantageous to make the commitment.

The implications for household energy in tropical countries are evident. Where sunlight is plentiful and annual temperature ranges above 5 degrees centigrade, PV systems may become the obvious first choice for newly industrialising economies. Their production as well as their application promises to be a growth-generative industry.

For business the implications are also great. British Petroleum, one of the largest oil producers, is also the second largest producer of PV cells. Within a few decades it may be a generic energy system producer, moving away from fossil fuel extraction and distribution as its major source of business.

In a few areas like Italy and Spain large solar power stations are being built during the 1990s. In Germany the '100 Roofs' programme was launched in 1990. Some 3,000 houses have now been fitted with solar electricity panels, with government contributing 70 per cent of their cost. The city of Aachen has imposed on electricity consumers a 1 per cent solar fund tax. This fund is then used to pay any company or household who exports solar power into the regional grid, at a highly profitable rate of 2 DMs per unit ($1.30). This market intervention is stimulating wide development of solar panels.

By 1995 some 35 Swiss and German cities had adopted the Aachen system. There is growing pressure for national legislation to reinforce the schemes and build up the momentum for solar power development.

Japan now has the most ambitious national programme, aiming for 60,000 new housing units per annum to be fitted with PV systems by the year 2000. In classic Japanese style, the government calculates that by guaranteeing through incentives a domestic market, new producers will come on line and produce PV systems that can become a leading export industry.

The United Kingdom government is waiting for technology to experience additional breakthroughs and price downwaves before putting government funding behind PV systems. Such organisations as Greenpeace are critical of this approach: the technology is available and with subsidies are cost-effective now. At Newcastle University a new office block has a front elevation covered in PV-based material; this provides about half the building's energy requirements and is thought to look and feel architecturally satisfactory.

In the coming age of environmentalised business, the housing industry will not be immune from the demands of technology. 'Smart' houses, houses that 'think' for themselves, will reinforce their earlier, pre-industrial role: centres for many forms of learning and productivity, with multi-household and multi-generational facilities built in. It is possible that the home/work environment realised by the new technology of the wired city and planet will support renewed focus on the family as a major social nexus.

Working all the time

Because of the ubiquitous nature of the Internet, it is not as simple to get away from the demands of work as it used to be. Instantaneous telecommunications are often characterised as a threat to leisure time, but it is also possible that the new professionals have no desire to 'get away from it all'. Newly emerging professional jobs as never before require a high level of commitment, enjoyment and hard work, forces which are now endemic to all professional and service industries.

In order to produce the steep rises in white-collar productivity required for success, most organisations have had to invest heavily in information systems and mobile technology. Wherever we are, we can connect with our work. If the bureaucracy of office politics is no longer a necessary component in order to produce high value for the customer and therefore for the organisation, work becomes more 'fun' and organic. Productivity need not be affected by the interpersonal politics which are so often endemic to working in close proximity to others.

Yet, loneliness and isolation is the central *angst* of modern life; hence the need for teamwork, for the consolidation of feeling and co-operation out of which production is improved. As the need for human connections accelerates with the centripetal pressure of advanced technology, the role of hotels, conference centres and of temporary office villages will increasingly displace the old permanent corporate headquarters pattern. Most of us will find that working from the house, at least some of the time, makes sense.

Teleworking and working: a false dichotomy

For many companies which operate 'hot desk' and 'no headquarters' policies, working from home is becoming not so much an option as an enforced mode of operation. The financial savings on real estate, made possible by wired and wireless networks, are so overwhelming that few organisations are going to be able to avoid the attractions of mobile working and dispersed workforces.

The home, therefore, becomes even more important, as the only certain work station which most of us will have. Space and facilities for work and document storage – we are a very long way from the paperless office – are essential in the professional homes.

A top executive of a large British home building firm told us in 1998, 'Today we cannot sell a house where there isn't room for a desk and storage cabinets. Even if they don't work at home yet, most people want the option of having these facilities should they require them.'

Breaking free of the car

With working from home, commuting to work becomes progressively optional. More critically, it becomes as the same time less fixed to the traditional 'peak' hours, morning and evening, five days a week, which have for the great part of the twentieth century structured our metropolitan regions. The resulting priorities have displaced mass transport and replaced it with massive road systems.

But the end is in sight. Even as it becomes possible to work from home and car as well as from the customer's workplace, it also becomes ever more expensive to travel into central cities. Governments threaten to tax car parking and the use of trunk roads during peak hours. Throughout Western societies and in such areas as Singapore, Japan and Hong Kong, the age of the car is still dominant, but the age of the car commuter may be coming to a close.

Mass transportation – fixed rail links, tramways, busways – may become a preferred investment option for twenty-first-century governments. But the car itself, as we see in other parts of this chapter, is changing its spots. 'Green' cars moving in more even flows across networks of freeways and not subject to peak congestion may ensure that the car in its new eco-behavioural mode remains the preferred method of moving from point to point.

Recycling human and household waste

Organic garbage recycling and human toilet waste recycling are well-established and indeed for much of the Western world already occur. Regional sewerage plants routinely use gases for power generation and process their intake into fertiliser components. Garbage is collected and incinerated routinely in combined heat and power plants in many cities.

In recent decades the mountains of garbage generated by affluence and convenience have become a major public ill. These accumulations will become recycled, reduced and reused. Increasingly the costs of incineration and landfill are being placed on consumers, to the degree that incentive for the three magic approaches (recycle, reduce and reuse) will be part of every municipal strategy. The business opportunity is to create products which are easy to recycle and reuse. Sturdiness and simplicity of construction will become as important as technological advance. The environmental agenda drives the design process and new housing design is adapting to this agenda.

Breaking out of the grid: water, sewage, natural gas and electricity

In early 1998, one of the worst ice storms in history affected eastern Canada and the north-eastern USA. Ice, blown by strong winds, clung to and froze onto overhead wiring across hundreds of miles, collapsing huge pylons in a disaster which had not been contemplated in the design phase of the grid which stretches from northern Quebec to New England.

For up to six weeks, vast areas of residential urban development were without power during extremely cold weather with below-freezing temperatures which lasted longer than normal. Tens of thousands of households were at risk and repair costs ran into hundreds of millions of dollars. Never has an incident of weather so crystallised the growing belief that dependence on the grids of electricity, natural gas, water and sewer presents an unnecessary risk and excessive cost structure. The technology of replacing the grid is already well established. It is the process of reinvestment and replacement which presents a huge business challenge for the next generation.

Essentially each house or neighbourhood can generate its own electricity through combined heat and power plants or through each house having its refrigerator-sized generator which converts natural gas into electricity. Again, the technologies for these quiet home generators are well established. But the business opportunity to make this an option has not yet arisen, in that the sunk costs of the grid operators allow them to continue to operate profitably in a period of falling home energy costs.

Garden design: rainwater recycling

The garden and lawns of the typical house consume enormous volumes of water and plant materials. Increasingly a new style of garden design is emerging which will allow the establishment and maintenance of home gardens without huge costs in plant food, fertilisers and excess water bills. In addition to built-in rain-water reservoirs coming off the roofs of houses, a range of drought tolerant planting, naturally equipped to cope with the range of variation in rainfall, is being articulated to offer an alternative to the labour-intensive traditional ornamental garden.

Passive solar energy: the new technology makes an old idea workable

Energy conservation has since the 1970s become a core focus for strategies across the industrial world. Energy efficiency has increased in some sectors by a factor of four since 1975.[9] With the East Asian recession of the turn of the century, this has meant that energy costs collapsed to the lowest level since the mid-1960s. The argument for energy conservation then resided in the aim to reduce greenhouse gases.

Over half of all atmospheric pollution arises from domestic heating. Therefore, with the advent of much greater efficiency in solar panel technologies from the 1990s, all this is changing. In some cases, home builders are installing solar heating as part of a basic package. Recently, a Swiss architectural firm called Architcam 4 has designed and produced a house that not only has modular walls which can be rearranged with a screwdriver depending on the size of the room desired, but the wall can then be folded and stored in the cellar. The houses are solar powered and insulated with sheep's wool and recycled newspaper fibre. Three of the houses have been built near Zurich and more are to be added.[10] In societies where the size of housing is largely driven by stages of life, the idea of a modular house which can reflect the needs of early marriage, children at home, children gone and retirement has enormous implications for the present approach which is to build houses for each of these stages, thus enormously inflating the demand for new housing.

Smart houses

In the UK the evolution of the 'smart' house becomes ever more approachable.[11] Structured with passive solar and heat pump technology, many homes are being wired so that a phone call from a mobile phone could check the house and start the process of preparing a meal. In some houses, up to thirty sockets are placed throughout the rooms, each wired with a high specification data cable linked to central controls. A range of appliances and safety devices can be installed and activated at a distance. Such homes also support multiple telecoms channels, with up to four Internet access gateways available simultaneously. Costs for this package are estimated at approximately $2,500 (£1,500).

Such packages are said to be for more affluent and mobile householders requiring maximum convenience. Yet for social housing operators, the 'smart' house option is becoming ever more attractive as a basic measure for lower-income households' needs. The Swale housing association in Kent in the UK, surveying its 9,000 tenants in 1998, discovered that the greatest satisfaction in housing service was having an immediate and professional response to a telephone call for repairs or emergency services. This function, Swale's CEO Stephen Howlett determined, could ideally be satisfied through a 'standard product' of the 1990s, the call centre manned by articulate telephone operators who would know how to get a rapid response from the housing assistance services as required.[12] It was only a short step further to identify a location on a nearby business park in north Kent at Sittingbourne and to establish a call centre. Most importantly its capabilities were to be marketed to other housing providers not only in southern England, but also nationally. Suddenly a housing services operator had plunged into the call-centre business.

Such 'intelligent' systems personalise communications between householders and environmental managers. Using such systems, better person-to-person relationships become fixed as part of the enhanced service available to households renting or owning houses in the Swale network. It is only a matter of time before such benefits can be extended to private householders seeking to add this support service to lifestyles requiring maximum mobility. This will be particularly true if, as planned, Swale is able to offer added features, e.g. free local telephone calls, a rarity in Europe in 1999.

Housing design: natural and environmental projects

Housing as a system is in the process of evolving more efficient and more effective users of community resources, using the technology of management and mobility to increase its functionality. Notwithstanding this technological future, the actual shape of houses is becoming more freeflowing and organic. In some cases houses can be built underground, a highly efficient use of land in dense urban landscapes. The desire for bicycle and pedestrian safety will increasingly mean that car speeds will be suppressed and traffic calming will dominate the suburbs where the greater majority of the population with small children reside. Example of efforts to achieve sustainable housing abound.

Ecologically responsible new towns

In Banberton, Canada, a new town is being planned whose essential ingredients will be community recycling, environmentally sound building materials, and energy-efficient designs and transport. It is owned by four large trade-union pension funds and is designed to house 12,000 people in the next twenty years.[13] One of the earliest examples in North America of a car-less, ecological town was 'Arcosanti', Paolo Soleri's experiment in the desert of Arizona.[14] One outgrowth of this experiment was the founding by Richard Register of Urban Ecology in 1975 in Berkeley, California, and a new organisation, Ecocity Builders, dedicated to the construction of demonstration building development and nature restoration. One of their earliest ideas was the concept of an 'integral neighbourhood', the construction of a village within a city.[15] In France, the revival of earthern architecture (based on adobe or sun-dried-earth brick) began in Grenoble and led to the creation of CRATerre (Centre de Recherche et d'Application Terre) which built 65 houses in the new city of L'Isle d'Abeau, between Lyon and Chamery, using rammed earth walls. CRATerre has now extended its expertise to projects in Africa and South America through the international network of Gaia, formed in 1983 by a group of architects, permaculturalists and biologists interested in an ecological approach to building and planning, particularly in the construction of the eco-cycle houses. In 1990 this influential group initiated Gaia International to form a network of ecological designers from 13 countries.[16] The Ecological Design Association (EDA), founded by David Pearson in London in 1990, is devoted to similar aims and has done much to foster public awareness of eco-design in the United Kingdom.

What is clear from the above examples is that the design of housing in a very short period of time has become increasingly conscious of the need for environmental sustainability and it is now an open challenge to developers to reap the benefits of that creativity.

THE ENVIRONMENTALISATION OF INDUSTRY

From many areas of development arising in industrial innovation, new environmentalised cars and buildings will increasingly be the centre focus for their respective industries in coming decades. They are being created in both the developing and developed world.[17]

The nations with the largest populations, China and India, both have large tropical areas with huge concentrations of population, urban regions with 30 million or more being common. As affluence increases, car acquisition and new residential and commercial buildings will form driving forces in these economies. At the same time, pollution risks and environmental nuisances will increase.

These countries have begun to attain growth success in their economies that has been persistently pursued over the last half century. It is already apparent that local and central governments will face further challenges in managing growth and focusing development in ways which do not compromise the gains being attained in the next half-century.

Predictably there will be a consequent intensification of environmental controls in the NICs in coming decades. Not only is there a conviction that toxic dumping is unacceptable: there is also the conviction that the most efficient technologies will also be more environmentally advanced. That is, cars which are fuel efficient are also recyclable and inexpensive to acquire. Houses, offices and factories which are cheap to build may be in a tropical climate inexpensive with regard to heat and light.

Technology transfers over coming decades within the global village will be far more efficient than anything heretofore. A hundred and fifty years ago on the American frontier, within five years of the first settlers arriving, churches, opera houses and elaborate restaurants, as well as forts, could be erected with kit systems transported by rail across the continent. Today, within 24 hours advanced-technology houses and cars can be transported any place in the world. Their technical specification can be faxed and wired instantly anywhere on the planet. As environmentally benign technologies expand, the top business people and government officials will be taking part of their professional training in European and American universities where the most advanced approaches are to be found.

It is unlikely that the values and expectations they develop in such settings will be abandoned as they return to their home bases and form their careers. Instead, it is as likely that they will attempt to compete with and supersede their advanced-economy classmates, in industrial, economic and environmental terms. This reality, despite the recent economic constraints induced by volatile financial markets, forms the inner story of, for instance, the tiger economies of South East Asia in the last two decades, as well as the takeoff economies we now see in Brazil, Chile, Mexico and other Latin American centres. At some point African nations will put behind them the tribal practices that have dogged their initial period of modernisation and find methods of marrying advanced-economy standards to cultural systems. Other feeder industries involved in the supply chain will be greatly impacted.

THE INDUSTRIAL METABOLIC ECOSYSTEM

The principle is becoming clear: industrial pollution signifies faulty processes and poor product design.[18] Engineers in industry all over the world are driving for comprehensive industrial ecosystems, by which factories located in close proximity to each other will feed off their mutual by-products and co-products. The target is jointly to accelerate profits and eradicate environmental costs. In time we will learn to decrease to near-zero unwanted by-products or 'waste'.

Creating new industrial metabolic enterprises requires a huge investment in human intelligence and engineering. It is in the environmental services industry, with its many firms and capacities, where this is emerging.

The sector has been expanding at a breakneck pace since the early 1980s, doubling in size about every three years. Many established international service businesses, including major accountancy and legal firms, began to form environmental divisions in this period. The doctrine of 'due diligence' is used by buyers of industrial and commercial assets. They must endeavour to discover potential negative liability in a business they are acquiring. This requirement has brought a great deal of commissioned work to these practices.

But new players have also been formed, using the values of environmental scientists and planners. In every country there are consultancy services driven by professionals trained in environmental engineering. These services are beginning to afford an alternative to the problems of making long-established companies more environmentally sensitive.

CASE: GREENING THE CHEMICAL INDUSTRY IN THE NORTH OF ENGLAND

No industry has had more to learn about its environmental performance than the chemical industry. Friends of the Earth, using official Environment Agency figures for 1996, created a league table in 1999 for the top polluting companies in England and Wales.[19] Associated Octel's plant in South Wirral appeared at the top of that list with 5,340 tonnes of carcinogens released into the air. In response to pressure from the EA, Octel's emissions have been reduced to 3,790 tonnes, but in 1996 it remained the top polluter, followed by ICI Runcorn with 2,151 tonnes and ranking well above Dow Chemical, Rhône-Poulenc Roroer and Ciba Geigy. Three ICI plants in Runcorn, Cheshire and two on Teesside were in the top twenty. Shell, featured in this book as a corporation which has become increasingly responsible on environmental issues, ranks 20th with only 37 tonnes. One of Europe's largest chemical companies is ICI. Its Tioxide plant in the north of England has been redesigning its process along environmental lines for half a decade. Tioxide is proving that, far from being a burden, environmental protection can be profitable.

Its manager, Rob Louw, drove the conversion of environmentally sensitive waste by-products into successful new co-products. Waste flows were converted from costs into a profit centre. Tioxide is Europe's biggest manufacturer of titanium dioxide. This is chiefly used as a white pigment for colourings in such businesses as textiles, building materials and inks. Between 1993 and 1995 sales of these co-products doubled from 400,000 to 850,000 tons. Remarkably, the sales value of these ex-waste products was in 1995 more than the value of sale of the main products of the plant.

The environmental problem stemmed from two main-by products, sulphuric acid and ferrous sulphate, known as iron salts. The company generated an annual total of 1.9 million tonnes. They used to be dumped either into the sea or into deep landfill sites.

From the early 1990s European-wide legislation required that these by-products be neutralised before dumping, in order to protect the seas and the land's underground water reserves. This requirement caused Tioxide to build a new factory which can treat the waste. As a consequence of the waste treatment, a building product called gypsum could be produced from a combination of sulphuric acid and limestone. The issue became whether this was to be dumped or to be sold to plasterboard and cement makers in the building industry. Not surprisingly, the decision to sell the product was taken. A new division, Materials Tioxide, was formed. By 1995 it was selling 30 products made from what had heretofore been waste streams; it included materials for fertilisers, building, water treatment, animal feed and carbon dioxide used in making bubbly drinks.

According to the business plan of 1996, by the year 2005 approximately 95 per cent of what had previously been dumped into the environment will have been turned into marketable products. One of the most interesting of those is Oasis, a water-treatment chemical in solid rather than liquid form. It is expected to find a global market and form part of the international effort in making available clean water in the developing world. So, not only has Tioxide stopped producing polluting by-products, but it has doubled its profits and demonstrated that the rationale for industrial greening leads to big benefits for business.

THE WORLD ENVIRONMENTAL SERVICES INDUSTRY: THE REVOLUTION LEARNS TO TURN A PROFIT

By the mid-1990s, the global market for environmental services was $250 billion, and it was growing at 20–30 per cent per annum. Countries with the strongest environmental legislation – USA, Germany, Japan – had to create an environmental services industry to deal with the environmental legislation and regulation created between 1970 and 1990. By the end of the century they monopolised this new engine for the creation of wealth. Other countries, like the UK, Finland, Sweden, Norway and the Netherlands, were rapidly creating smaller firms. These were clustering together in design and research on environmental affairs and were able to compete internationally.

By the late 1990s, environmental industries were in a critical stage of development.[20] Rapid growth in all countries triggered mergers, acquisition and other restructuring in the emergent sector. In general, environmental services is ideal for smaller firms, because they can respond flexibly and with innovation. However, internationalisation of the sector has made small firms in other ways less than ideal.

THE ESTABLISHMENT BUYS INTO THE ENVIRONMENTAL REVOLUTION

For most manufacturing industries, attaining environmentalised products and processes is an important step in maintaining and taking market shares globally. For instance, chemicals, cars and steel sectors have all had to respond to both governmental regulations and international export markets. To do so, these sectors have turned to outside sources for advice and technology. Thus, the growth of the environmental services – including technological advice – sector has been rapid, approaching by some estimates 30 per cent per annum globally.

Smaller firms drawing on specialist scientists, engineers, and management consultants have in general been in the lead. Firms with long-standing environmental interests, such as city planners, civil engineering and architectural firms, have also rapidly developed environmental sciences and engineering interests. Most remarkably, firms that have been traditionally identified with 'hard-headed' approaches to business, law and accounting, have found that their major corporate clients had all developed substantial appetites for environmental assistance. The need was to avoid legal and product liability, conform to legal regulatory regimes and most of all to protect corporate value.

By the 1990s, corporations at the heart of the world's largest investment markets, including accounting firms (Touche de Loitte, Kate, Peat Marwick) and law firms (McKenna & Co.), had environmental sections offering specialist advice to their clients. While strategic consultancies such as McKinseys and Bain & Co. had not assembled specialist teams themselves, they formed working alliances with leading specialists such as ERL, Ltd, A.D. Little Inc. and other major operators. It is notoriously difficult in a highly fragmented industry like environmental services to estimate global market size in this sector. However, it is safe to say that during the 1990s business was probably paying out yearly amounts approaching $500 million for environmental advice. By comparison with mainstream billings in strategic, legal and financial affairs, this amount is not great. Yet this sector of advice hardly existed as recently as 1984.

SUMMARY: STRATEGIC TECHNOLOGY-LINKED CORPORATE AND BUSINESS CHANGES CONTRIBUTING TO ENVIRONMENTAL SUSTAINABILITY

By the late 1990s, world business was effectively assimilating the environmental agenda into its strategic framework. The corporation is pushing into place many strategic systems which have environmental consequences, as the following trend survey illustrates.

Use of homeworking and telecommuting

Allowing increasing numbers of workers to work from home will, while reducing demands on core offices, also decrease peak flows on transport and road corridors. Over time, net commuting demands may well be smoothed out over the working day.

Downsizing and outsourcing

Smaller headquarters staff means many smaller businesses will be asked to provide key services and products to service industries. Because this system is inherently more widely spread over geographic areas, there will be benefits for sustainable regional systems, with less congestion and greater use of telecommunications.

Decentralising

The city continues to offer much to the senior levels of large and smaller businesses. However, many routine operations can be placed away from large metropolitan centres, with direct benefits on staff retention and availability. There are also environmental benefits in reviving outlying provincial centres and creating new areas of employment growth.

Developing energy management systems

Energy management has obvious benefits for decreasing business costs. However, no greater direct benefit to sustainability can be offered than for businesses to see energy saving, conservation and management as a prime area of environmental development.

Environmental specification for new building

As recently as the economic boom of the 1980s, environmental sustainability in architecture and town planning was regarded as a side issue to aesthetics and profit. By the late 1990s, architects and civil engineers world-wide were fully focused on new buildings as prime areas of contribution for environmental conservation. Many real-estate firms and architects made environmental conservation a prime part of their commercial marketing to business clients.

Environmental retro-fitting of production and space facilities

Old buildings are also increasingly being redesigned to save energy. Factories, to conform with regulation, were having to be rebuilt. Many countries provided special subsidies for improved insulation of residential and commercial premises. More importantly, energy producers were given financial incentives and complex telecommunications systems to enable large users to decrease and control demand during periods of intense usage.

Use of telecommunications and real-time management of distribution and logistics systems

Video-teleconferencing, by the late 1990s had become commonplace. As earlier forecast, it was able to replace the need for international movement of company executives, in many instances. But more importantly it demonstrated the power of telecommunications. Thereby, a ready acceptance of many forms of logistics control and communications was put in place.

9 Nations and Cities: Sustainability Models

Reality Field 9
The innovative business and the creative city region require one another.

There is a comprehensive interrelationship between excellence in government planning and long-term wealth creation. In building cities, business mediates among communities, families and the planetary environment. Corporations need to aim to design environmental systems which are responsive and re-programmable. By so doing they offer the most successful and tested approach to cities. Only with business resources will cities be capable of absorbing the expanding human population.

THE END OF URBAN DECLINE

As was reflected in the proposed regeneration of Central London's Kings Cross (Chapter 6), the reputation of the modern city in the twentieth century, long declining, is now the target for reversal. With the continuing increase in environmental problems and human population world-wide, government and businesses are in the midst of reinventing the modern city. In order to accommodate an increasing population and aspirations for better environment and economic prospects, it is becoming clear that the city in all its forms and sizes is not the problem but rather the solution.[1]

The problem of the city has been around for centuries and it intensified with initiation of the industrial period of economic development. About three-quarters of the human population in the late twentieth century live in countries which are newly industrialised or hoping to become so. In 1952 there were two mega-cities in the world, that is two cities with a population greater than 8 million. In 1997 there were 25.[2] It is in cities where this industrialisation takes place. Therefore, we can expect that cities in these countries will be the focus of opportunities and problems – social, economic and environmental – over coming decades.

The cities of America, the tiger economies of East Asia, and Europe can be expected to continue their transition into information-based dispersed urban regions. There will remain pockets of social deprivation, particularly in the USA with its long history of enormous ethnic migrations and lingering racial tension. Much smaller but not dissimilar pockets of ethnic concentration, poverty and racial tension are found in contemporary European cities – Marseilles, Paris, Liverpool, London, Dresden, Halle. These communities in the advanced economies will continue to be a declining, substantial and minority component of the overall urban pattern.

The great majority of working-and middle-class populations will increasingly choose to live in less dense outer urban centres. Two-income families have become increasingly prevalent, as women become an ever higher proportion of the workforce. Housing will be chosen which offers a range of employment and personal development opportunities. More than ever before access to good schools, universities, job opportunities, shopping convenience and leisure options in the countryside will become part of choosing houses. Existing pressure on housing and schools will be intensified as more and more people move into the salary ranges and consumer habits of the middle class, however problematic it may be to identify that class. For example, if class in the UK is defined largely by occupation, it can be demonstrated that the middle class there, as in the rest of Europe, is growing:

- A – Upper Middle Class: Higher managerial, administrative or professional
- B – Middle Class: Intermediate managerial, administrative or professional
- C-1 – Lower Middle Class: Supervisory or clerical and junior managerial, administrative or professional
- C-2 – Skilled working Class: Skilled manual workers
- D – Working Class: Semi-and unskilled manual workers
- E – Lowest Subsistence Level: State pensioners or widows (no other earner), casual or lowest-grade workers.

Based on the new occupation hierarchy, recent survey data suggests that the middle class is slowly expanding, although the percentage has also increased for Category E (Table 9.1).

Table 9.1 UK social stratification, 1975 and 1997–8 (percentage of total population)

Social grade	*1975*	*1997–8*
A – Upper middle class	2.3	2.8
B – Middle class	10.4	18.0
C-1 – Lower middle class	23.5	27.5
C-2 – Skilled working class	32.5	22.0
D – Working class	22.0	18.1
E – Lowest subsistence level	9.5	12.0

Source: National Readership Survey (see note 2).

As the middle class expands, pressure for new housing intensifies, particularly in densely populated countries such as Great Britain. Government, caught between the demand from environmentalists that the open countryside should be preserved and the demand for additional housing, tries to find a compromise which increasingly focuses on existing building sites within cities, especially the so-called 'brownfield sites'. Emphasis on city regeneration may well be a solution to the demand for housing which is compatible with a desire to preserve green space, but it will require innovation and creativity from developers, housing corporations and governments, especially with respect to the housing needs of those living at the lowest subsistence levels.

By the last decade of the twentieth century, London, New York and Paris have been able to stabilise the decline of their populations. As noted in the previous chapter, New York is seeing for the first time in many decades falling crime figures. Changes in policing management, heavily influenced by business corporation practices, incresingly suggest that community involvement and preventative strategies which are targeted at vulnerable groups and locations will reduce urban disorder. This increasing community control is consonant with retaining a mixed middle-income population. And these populations are necessary to ensure the continued quality of educational and health care resources.

URBAN INFRASTRUCTURE AND SUSTAINABILITY

The challenge of rebuilding cities over coming decades is already being addressed in many parts of the world. In London, the Ebbsfleet

New Town development in Kent county on the Euro Tunnel Link to Paris is being designed from the ground up as a low-energy high-technology urban complex. In general, Japanese and European cities already have high levels of public transport. The strengthening of public systems is a constant focus of investment in Amsterdam, Paris, Berlin and Hamburg.

But North American cities are changing too. As a result of the last quarter century of civic innovation, they are well on the way towards making public transportation a primary method of moving people across urban regions with many different centres. Los Angeles' recent metro system is a grid connecting many centres in this spreading metropolis, and Washington DC has a new underground which is a model of clean, quiet transport. Toronto and Montreal also have well-established and efficient systems of underground subways. The New York subway system, one of the world's oldest, has been partially renovated in recent years and forms part of the restoration of civic pride to one of the world's great cities. Boston and Philadelphia, Cleveland and Chicago are all experiencing revitalisation and increasing reliance on public transport in recent years.

It is in some of the less well known cities where notable progress is being made. The cities of the Mid-West and Far West are interesting because they grew up in direct response to the car and never had the higher densities or single cores which railway-age cities of the nineteenth century most often feature.

A case in point is Portland, the principal city of the north-west state of Oregon.[3] A traditional port and highway hub, Portland was festooned with freeways by 1980. Under the leadership of Governor Tom McCall and Mayor, later Governor, Neil Goldschmidt, Portland over a twenty-year period undertook to build in a public transport preference for its central city development. Using the land use approach to American development, zoning laws were rewritten to discourage parking. Instead of specifying minimal numbers of car parking spaces, maxima were set. In one case a six-lane freeway along the Willamette River was torn up and replaced by an extensive urban park. This park provided the amenity setting around which a large range of new office and residential developments was generated. This in turn brought a renovation of investment interest in the core area. A light rail transport system connected the revitalised urban core with rapidly growing suburbs. Based on these changes and its good government and community relationships, by the early 1990s Portland was voted the most desirable American city in which to live. Drawing upon the experiences in Portland and other American cities, what is required to achieve urban sustainability? Reducing consumption seems to provide a good beginning.

Sustainability means reducing consumption

Bringing demand for infrastructure improvements under firmer policy guidance is the core issue.[4] Total vehicle travel in America has grown 5 times over the 40 years 1955–95. There is now twice as much solid waste per capita produced as in 1955, and the population of the USA has grown by 60 per cent. In many cities, water consumption is climbing at 5 per cent per annum as population grows at 3 per cent or more.

In general, consumers are shielded from the costs of growing urban infrastructure. Beneficiaries of infrastructure improvements are not charged the full cost of their development. Instead, the load is spread across the whole of the population's tax contributions. This makes sense if the infrastructure development promises to produce public efficiencies and environmental benefits which everyone will share in some proportion. However, in the case of highways and the great infrastructure of car movement, car and truck users in the USA pay only a small proportion of this cost.

In many European countries, there are now calculations which show that the proportion of car subsidy is much lower, but the purchase price is much higher and varies widely within Europe itself; for example, by 1999 the Netherlands was the cheapest market for automobiles and the UK was the most expensive. However, in no country are the full costs of motoring imposed on car users, apart from Singapore where a road pricing system is now in use on some routes within this small country.

What is needed are better stick-and-carrot methods. Government changes should be designed so that in a systematic and incremental fashion, not imposing too many 'market' dislocations, a preference for sustainable options can be built into infrastructure decisions. The case of energy consumption is one which provides some hope.

The reality of 'soft' sustainability achievement: the case of energy

In his research and publications of the 1970s, the energy scientist Amory Lovins first showed how a policy of sustainable options might work. He outlined how super-efficient technologies – from light bulbs to electric motors to industrial refrigerators – would lead us down a 'soft' energy path. He forecast that electricity use would be dramatically cut while living standards would be maintained. In the last twenty years this projection in large part has come true.

By the 1990s, twenty of America's top fifty energy producers provided direct incentives to their consumers to reduce energy use. In the face of a doubling of GNP since the mid-1970s, the demand for electricity has levelled out across the USA and is now roughly what it was in the late 1970s. Equipment has become more efficient. Service levels and standards of living can be maintained on far fewer kilowatts. The most expensive, dangerous and polluting power plants can be brought out of service, as they are replaced by 'cleanburn' plants.

An important fact is that small investments in new technologies have produced huge savings across society. Lovins's concept of the 'negawatt' measures electricity supply as a function of this efficiency. Power producers in California, like Southern California Edison, have learned to profit from producing less electricity. Between 1981 and 1994 electricity demand was cut by an average of almost 500 megawatts a year, equivalent to 3 power plants. Profits grew, since expenses – in terms of capital investment and operating costs – fell faster than revenues.

Policy innovators now see the possibility for extending this principle. In the same way that pricing has produced a decrease in electricity demand per unit of wealth generation, so it can be used to decrease environmental costs by setting up systems which reward households and firms for:

- negative miles travelled by car, or negamiles;
- negative gallons of gasoline consumed, or negagallons;
- negative pounds/kilograms of household solid waste produced; or negagarbage.

Utility operators can be required to carry out programmes to provide incentives for reduced use of their services. To do so companies will have to quantify usage and implement strategies to moderate consumption. Financial return is distributed to customers within the region. The utility corporations can issue new debt for this approach just as existing utilities do. But as with the power-reducing generators, they will use incentives to cut consumption levels through investment in such areas as water-saving technologies, public transportation, telecommuting and waste reduction. Clark Wieman, director of a

New York City public policy think tank, observes, 'Indeed, the new negaunits, like negawatts, become a commodity that can be traded between utilities, even between customers.'[5]

CASE: LONDON'S GATEWAY REGION: A KNOWLEDGE CITY COMING INTO BEING

MetroCity-on-Thames

In the Docklands area and Thames estuary reaching east from London, a much-interrupted series of developments have occurred in the 100 square miles. Between 1975 and 1995 some £5 billion/$8 billion of private and public investment occurred. In the inner area of Docklands, population has gone from 50,000 in 1970 to some 120,000 in the mid-1990s, as large areas of waterfront have been redeveloped for middle-income housing. The majority of London's international and national newspaper publishing activity has gone east. A number of international banks and investment houses are moving into the marble-and-glass palaces towering over the River Thames.

A new commercial and financial district centres on the skyscraper of Canary Wharf, which has extended the central business districts of London to the east for the first time in two centuries. These developments have transformed the economy and society of East London, as will the new Millennium Dome currently being completed. Further down the Thames, the opening of the Euro Tunnel in 1995 has increasingly made possible a transformation of worn-out urban regions which have suffered neglect. On the eastern edge of London, a new international rail station near Dartford in Kent will create a new town setting which will substantially pull regional development from the west of the region towards the east.

The key method for this development has been a series of joint venture corporations. In the Docklands area of east London, the London Docklands Development Corporation during the 1980s became the major public planning and development centre, taking control from the local municipalities which had been unable during the recessionary climate of the 1970s to come to a consensus about how development should proceed.[6]

Susan Brownhill adopts a critical perspective in referring to Docklands as another 'great planning disaster'. Her analysis focuses on the social costs imposed by urban corporations in developing Docklands.

The social costs have undoubtedly been great, falling on the 10,000 residents immediately affected by Canary Wharf. With low education and job skills, few of them have been able to find work or to buy the new houses in the Docklands development process and have been displaced.[7]

The counter-argument centres on the benefits of the development, which relate to the whole of the British economy. Major export sectors, especially for London, are tourism (ranked number 1) and financial services (number 2). Docklands, it is now apparent, will contribute substantially to the competitive advantage of the British economy in regard to financial services by lowering the rental costs required of overseas banks locating to Britain. In recent years the new centre at Canary Wharf has also become a focus for tourism and the whole of East London's architectural heritage has been reinvigorated and discovered. These structures have become marketed as secondary tourism targets.

THE URBAN ADVANTAGE

Why will business firms hold on to their urban bases and thereby sustain economic viability of urban regions in all countries? The reasons are simple. The revolution in telecommunications and wireless information transfer will not displace but will supplement existing modes of business marketing and logistics. The wired city will be planetary in extent and the appetite for close proximity and multiple channels of serendipitous communications will ensure that cities flourish.

Like firms, cities must have competitive advantage against one another and with non-urban locations. Companies in cities see more of their customers and suppliers than is possible with even the most intimate, inexpensive and user-friendly telecommunications. Cities offer many opportunities for serendipitous – accidental – marketing. They comprise for all their residents 'opportunity clusters', where close proximity and both random and planned interaction impact on each other continuously.

As society leaves behind the industrial age, the single-industry cities of the past will become diversified or decline. This happened in Detroit where cars have moved away, and in Pittsburgh with the death of big steel. A similar decline in manufacturing and its reverberating effects can be seen in Sheffield, Manchester and Leeds. With the hollowing-out of the old factory and office monoliths, zones of entrepreneurial choice are being created. However, to make them forge ahead, a wide agenda of government and business catalysts are needed. In some cases, even major commitments by companies and government as in Detroit will not be enough to reverse the decline quickly.

Pittsburgh, for example, is located in a rapidly diversifying regional economy which is recovering from the loss of steel. Milwaukee, Boston and Philadelphia are other major urban regions which have successfully transited from industrial to knowledge-based regional economies, due in large part to a great proliferation of university research and development.

CASE: THE DUTCH RANDSTAD: AN INTERNATIONALISED RIM CITY EVOLVES

Nowhere are the possibilities of the highly dense information-based internationalised urban economy better expressed than in the great rim city of the Netherlands.[8] This comprises a circle of principal cities surrounding a green reserve of lakes, woods, agricultural conservation zones and nature centres. Its population of about 7 million includes Amsterdam, The Hague, Rotterdam and Utrecht.

With 650 persons per square kilometre, its population lives at a density double that of urbanised Japan, and somewhat greater than that of Greater London/Southeast England, the other major West European megalopolis. In spite of high labour costs and social taxation, the Netherlands has been successful at bringing in overseas investors – 20 per cent of recent USA investment in Europe, and, as well, many British, French and German concerns. The Randstad has provided many of the preferred locations for this investment.

It offers three main advantages which overcome its high land and labour costs as well as its high taxes. First, it offers immediate access to the whole of the German economy, upstream via the Rhine which flows into the sea at Rotterdam. Second, it has unexcelled air, rail, sea and motorway transportation networks, reaching all major European cities within three hours as well as offering one of the most efficient of international airports at Schipol.

Finally, and less materially, it offers a climate of individualism and free enterprise, combining Germanic efficiency with a preference for the English language. For these reasons, the Randstad for the first time offered serious competition to the United Kingdom and London for inward investors from Japan and North America, as well as the rest of the world. Whereas German cities offer little English-language support, Dutch cities work happily in the most successful international business language. Unlike France, there is little overt commercial protectionism and, some say, an absence of chauvinism. In comparison to British decision-making, Dutch municipal corporations and investment agencies have developed a strong American-style rapid-action Can Do culture which deservedly wins business.

The main Dutch universities and many of the secondary schools use English as a principal instructional language, recruiting top academics from around the world. Coupled with a vigorous musical and artistic life, the Dutch Randstad offers one of the most competitive city states in Europe. In that specific area of knowledge-society activity, software, Holland is attracting many software entrepreneurs from across Europe who like its easy access to the whole of the continent without the stifling cultural and bureaucratic constraints of England, France and Germany. Such innovations as the sheltered Centre Parc resorts, offering mid-winter tropical swimming in a forest environment, are Dutch developments being exported to Britain and Germany.

The secret of competent governance as in other places is that government structures and urban planning processes are constantly undergoing review. In the mid-1990s the Rotterdam regional government is being taken apart, decentralised and then re-aggregated over a much wider region in order to respond to changes in the economy and social structure.

THE UNIVERSITY AS A REGIONAL KNOWLEDGE FARM, FACTORY AND NETWORK

In the Netherlands and in every other country, the knowledge-based city is one where knowledge companies that are small and fast-moving thrive. Large companies are increasingly selling off their non-core businesses to their managers. A multitude of smaller businesses are arising in the place of a small group of large vertically integrated monoliths.

These smaller firms will require many methods of gaining efficiencies previously offered by the large corporation. The regional university with a wide range of part-time and joint-venture structures is a key method of facilitating urban change and business development, both in older cities and the expanding suburbs.

In the London region, the regional economy extends far beyond boundaries of the ring motorway encircling the traditional built-up area. With major greenbelt zones safeguarded immediately beyond the older suburbs, the London economy extends at low densities interconnected with motorway and rail networks across a field of some 20 million people stretching 250 kilometres from the traditional centre.

Within this area some 120 universities and colleges have emerged since the middle of the century. Many of them started as polytechnics and colleges, with highly vocational and professional missions and a wide range of part-time and evening courses. As universities with government charters and subsidies, they continue this emphasis even as they acquire some of the traditional liberal arts and high-science skills of the traditional university.

Similar expansion of institutions and regional networks is characteristic of, for example, New York, Paris, Toronto, Milan, Zurich, Hamburg and Munich, as well as the American cities already mentioned. These non-centralised urban networks are becoming in turn the model of urban development in Bombay, Mexico City, Jakarta, Manila, and Nairobi.

SMALLER BUSINESS AND THE URBAN ENVIRONMENT

The question to be posed is how business and the environment relate to this pattern of development. Businesses that are decentralised, smaller and fast-moving will increasingly replace material consumption with knowledge-based skills, where the value-added component and the price of skill are higher. A software firm is able to charge higher premiums for its products than a traditional book publisher. And the best software firms' ideal size – always excepting the monolithic Microsoft – may be much smaller than many publishing firms.

These firms will occupy smaller spaces; many of their key people may prefer to work from home or from local micro-offices. Employment itself will increasingly be replaced by sub-contractual relationships for particular projects. As contractors move from one firm to another, they will need to be in a regional urban network offering many such contracts.

Increasingly large offices and factories are being broken into smaller clusters of short-let serviced offices and workspaces. Thus while large corporations may become the exception rather than the rule, the industries which they related to in the past will continue to characterise the city regions where many new firms are emerging. Los Angeles and Hollywood, in the post-major-studio era, remain the centre of TV and motion picture production on a world scale. Boston remains an insurance and financial services centre even as it adds a wide range of high-technology companies to its suburban economies.

New York has many different economies, but remains a centre of clothing production and fashion design, as do London, Paris and Milan. The clothing and design sector has always had many smaller firms rather than a few large ones. Because of the overwhelming influence of professional skill, this industry represents a continuing model of change and adaptability which other industries, such as financial services, automobile manufacturing, heavy and light engineering, are all having to accommodate. Software, a recent explosive growth industry at the edges of New York City's garment district, provides yet another example of successful smaller businesses.[9]

Although the merger mania so characteristic of the late 1990s indicates a trend toward consolidation and the creation of ever-larger entities in certain sectors such as aerospace and defence, it has the opposite effect of reducing the number of companies overall and may well provide a stimulus to the creation of smaller, supply-chain businesses.

PLANNING THE SUSTAINABLE NATIONAL STATE

While corporations, cities and regions are all undertaking efforts to make themselves sustainable, whole nations as yet are slow to adopt sustainability as anything but a long-term goal. The exceptions are some of the smaller, in the sense of more compact, countries. Sweden, Denmark, Finland, and Norway have all adopted national policies which will produce contributions to lowering atmospheric carbon loading.

Netherlands, with its 20 million population on a minute land area, is exceptional again. Here, with the agreement of a national consensus embracing farm and city, large company and small, city dweller and cosmopolitan, the country has accelerated since the 1970s a national process of establishing alternatives to the car. Public transport is pervasive and deeply subsidised. Building insulation is a national obsession and energy sources have been rebuilt to use waste.

All is not plain sailing: A major source of ground and water pollution are animal agricultural wastes. In spite of intensive research, recycling and reuse of animal wastes has not turned out to be economically viable. As in Germany, packaging laws in conformity with European Union indications have produced dramatic reductions in commercial waste.

Some of these experiences can be seen in federal states, like Germany, Canada and the USA. In specific regions and localities, initiatives and a strong political consensus are producing great changes in environmental performance. In California, Pacific Gas & Electric, the monopoly provider of energy, provides lower-income customers with free house insulation, low-energy light bulbs and other incentives. Its own regulator is the source of the pricing regime which rewards the energy supplier by allowing higher prices in a regime of energy-use reduction.

SUSTAINABILITY IN NEWLY INDUSTRIALISED COUNTRIES: COSTA RICA

In 1995, about half the human population lived in cities. In the West this proportion approached three-quarters. In developing nations, the city continues to offer the best prospects to rural dwellers where overpopulation and land shortages predominate. By the year 2000, 17 of the 21 major cities with more than 10 million people will be in developing countries. Yet only half of the population in newly industrialised countries (NICs) is presently urbanised. For business and government the implications for urban management are great. With much lower incomes, the NIC urban immigrants are faced with major challenges.

But more telling for the future are lower-income countries still going through development and rebuilding. These countries are the greatest potential source of increased pollution in coming decades. If in a developing country setting, sustainability can be established alongside economic growth, then the larger challenges, India, China and Russia, may succeed in adopting parallel practices.

Such a laboratory case is Costa Rica, on the west coast of Central America. Here in the last half-decade and led from the presidency, a small, less-developed country with strong commitments toward sustainable policies in agriculture, industry, and urban development has emerged.

CASE: COSTA RICA: THE GREENING OF PARADISE

From his election to the presidency in 1994, Costa Rican president José Maria Figueres has committed his government to a policy of attaining for Costa Rica a sustainable economy.[10] Whereas many countries during the post Rio 1992 period have attempted to advance the cause of the environment, there have been few countries which have actually pursued the commitment as a national priority.

Costa Rica with a population of 3.34 million, restricted land area in a Central American region of great scenic and agricultural value, and long tradition of peaceful change, constitutes an ideal developing country in which to experiment toward a sustainable development process. In addition, its political culture includes an active entrepreneurial middle class, a free press, universal education and an egalitarian functioning democracy. In 1948 the present President's father, President Don Pepe Figueres, abolished the armed forces. Thus Costa Rica has been spared the military take-overs that have plagued the rest of Latin America.

Costa Rica has severe restrictions on its land resources. Its population growth rate remained in 1993, at 2.38 per cent annually, much higher than the global average of 1.74 per cent. Until recently, it was allowing deforestation to proceed at a calamitous rate, peaking at 100,000 acres per year. This had been reduced by 1995 to 20,000 acres, due to the establishment of parks and reserves. Much of the deforestation had been to extend cattle grazing. As world beef prices have slumped internationally in the 1990s, most of the cleared land has become idle and should be appropriate for reforestation.

Figueres, educated at the American military academy, West Point, has focused on four areas of policy development: taxation, energy, education, and national parks and reserves (see Table 9.2). Working with a national task force, he maintained that economic development will be accelerated through environmentally sustainable development policies. He identified two areas of international rapid growth requiring high-quality environmental conditions to attract investment: tourism and telecommunications development and manufacturing.

Most of all, Figueres' own leadership appears to be crucial. Each of his cabinet ministers whose portfolios include environmentally linked responsibilities has been required to prepare a strategic plan showing how public policies are going to be progressed toward sustainability. His national framework includes four major fora per year on the

Table 9.2 Costa Rican initiatives to establish sustainability as a national priority, 1994–6

Programme area	*Measures*
Taxation	• Establishment of carbon tax to help fund restoration of forests on vacant grazing lands • Electricity tax to encourage conservation and to help fund tree planting on forest slopes around hydroelectric dam impoundments
Energy management	• Additional taxes on fuel and electricity consumption • Opted for geothermal electricity generator, over oil-fired, which will by 2000 produce 25% of electricity • Required all home appliances to be 'energy conserving', through tax credits for low users • Established a goal of producing 98% of electricity from renewable sources by 2000
Educational programmes	• High-school pupils have annual extended field trips in forest habitats to focus on needs and potentials of eco-development of national habitat • National programme of conferences and community forums involving local and national leaders in quarterly meetings on national strategy
National parks, zones of forest conservation, and nature reserves	• Establishment of 2 new national parks, one to be funded by USA's National Fish and Wildlife Foundation • Extend conservation areas to encompass 25% of national land area. • Restricted development of Pacific coast resort hotel development to size and scale in keeping with ecological constraints • Redirected development of paper mill and port to less sensitive coastline and ensured good planning and government co-operation to expedite development • Undertaken measures to increase and conserve biodiversity throughout national ecosystems • Established national ecological research institute to link biodiversity, forests and strategies

environment and development, involving several hundred trade-union leaders, industrialists, civil servants and educators.

Costa Rica has become, through this process, one of the first countries to institutionalise the findings of the 1987 World Commission on Environment and Development. Named after Gro Harlem Brundtland, the Commission's report underscored the point that is being implemented in Costa Rica: development cannot be sustained for long if industrialisation destroys soil, forests, water and air resources.

In order to reverse the problems of overfishing, monoculture in extensive banana plantations, and deforestation, the measures undertaken appear to be almost too late. A World Resources Institute study of 1991 calculated that in the prior two decades, environmental resources with a value of $4.1 billion, some 23 per cent of the national capital resource, had been sacrificed to unsustainable development.

National parks therefore are being developed as economically active landscapes which, while sustaining biodiversity, natural habitats and conservation values, also pay their way in the national economy. The industries of eco-pharmaceutical harvesting and tourism are appropriate in these areas. Eco-tourism is an enormously successful strategy and the sector is booming as more and more travellers opt for Costa Rica's rain forest. The 55,000-acre Children's Eternal Rainforest, run by the Monteverde Conservation League, and adjacent to the Monteverde Cloud Forest Preserve, is indicative of the imagination and innovation which has been brought to the ecological problems of Costa Rica. Pharmaceutical companies have become increasingly aware of the potential for naturally sourced plant materials in innovating new health-care remedies. Costa Rica's remarkable rain forests with an inventory of an estimated 333,000 species, constitute an important concentrated landscape capable of considerable contributions in this new eco-technology. Merck Inc., one of the largest pharmaceutical multinationals, is involved in assisting in research and innovation studies within the sustainability framework.

In another cross-national initiative, a Nebraska power producer, Tenaska Inc., has funded an 8,000-acre reserve, previously destined for cutting, in order to satisfy its American-imposed contractual obligation with the Bonneville Power Authority to offset greenhouse gas emissions with conservation measures and 'carbon storage' initiatives. There is also considerable research on bamboo trees as a fast-growing high-quality construction timber for tropical house building. The Dutch government is providing some $6 million of funding for this project.

Costa Rica as a model for other countries

By the late 1990s the Costa Rican model, at first dismissed as a small country with special circumstances, is providing best-practice examples not only for other developing countries but also for countries with advanced economies. In Wisconsin, foresters, looking at the benefits of rationalising clear-cut strategies to protect old-growth forests, have accepted the need to alter forest cutting along the Costa Rican model. The bio-institute, INBio, has become a world leader in nationally controlled innovative eco-technology and is being emulated in, among others, Indonesia, Mexico, the Philippines and Kenya.

While all these measures look successful, Costa Rica remains a poor country, with $2,000 per capita GNP, as against a common European/American figure of $20,000. That said, many local groups are emerging in Costa Rica and taking responsibility for local watersheds and for the environment. As other countries have done, Costa Rica has realised that there is greater wealth in tourism than in export of commodity-level food exports. Thus, in a number of locations, restoration of natural landscapes in the rich volcanic highlands is displacing worked-out banana and coffee plantations. Within the new national parks, a wide array of outdoor recreational pursuits is on offer to international tourism, including wildlife expeditions, river-rafting, canoeing, mountaineering and hiking.

Costa Rica does indeed have advantages in biodiversity, tourism, democracy and education. But Daniel Janzen, a University of Pennsylvania biologist who has lived in Costa Rica since 1963 and assisted many Costa Rican projects, argues it is important that national eco-planning be made to work in just such an 'easy' country: 'The Wright Brothers didn't try to fly to the moon. But they did demonstrate that humans could fly. This is what Costa Rica gives us.'[11] More recently he and other biologists in Costa Rica are focusing on the preservation of biodiversity as the central purpose of preserving the dry forests of Costa Rica, particularly in the 323,600-acre Guanacaste Conservation Area, a mixture of parks and recovering ranch land. Jansen and his wife, Winnie Hallwachs, helped Costa Ricans raise $28 million to purchase the lands, established a 120-member staff and set up a $12 million management endowment. The conservation area now includes a tract extending 12 miles out into the Pacific Ocean and stretching to the Atlantic foothills of volcanic peaks 40 miles inland.[12]

Similar examples of developing countries preserving old-growth forests, safeguarding biodiversity and attempting to reverse the negative effects of indiscriminate tree cutting may be found in the Philippines' Haribon Foundation. The Foundation is dedicated to protecting the Isarog forest, where between 1961 and 1988 more than 40 per cent of the bird species had vanished. Although much of the lowland forest has disappeared, the cutting has been stopped and the villagers work as volunteer guards to prevent illegal cutting. Unfortunately it required a mudslide in 1991 which killed 7,000 people to force local and national realisation that stripping trees and undergrowth with the resulting loss of soil on mountain slopes will lead to disaster. In 1998 similar mudslides occurring in the midst of a hurricane in Nicaragua and Honduras had the same predictable results.

A CLOSING NOTE: ECO-DEVELOPMENT AND THE DISPLACEMENT OF INDIGENOUS CULTURES AND ENVIRONMENTS

There is little doubt that, however visionary and socially aspiring eco-development in developing countries and developed cities is, there will be negative impacts on many communities and individuals. No better example exists than in London where inner London is now the wealthiest single region in Europe, with a per capita GDP more than double the European average and three times higher than the poorest British regions such as Merseyside, Cornwall and West Wales, thus making Britain the country with the sharpest distinctions between rich and poor regions.[13] Some of outer London ranks nearly low enough to qualify for special aid from the European Union. But the argument that displacement should be avoided must be placed against the counter-argument that these forms of development, however difficult, offer a positive and manageable alternative to the only other approach, free-wheeling capitalism driven either by multinational companies, irresponsible national firms or by the failure of national governments to provide plans and controls.

In the coming century no human community will happily reject the new global economy and all its benefits. In this respect, therefore, managed and eco-focused national development plans offer the best workable approach for every developing country.

Part III

Ethics and Futures: Social and Environmental Ethics

Although the concept of environmental ethics is pervasive throughout this book, in Chapters 10 and 11 it is presented as part of a much broader perspective that encompasses the social dimension. After an introductory survey of teaching ethics in business, Chapter 10 discusses the national, international and meta-theoretic dimensions of business ethics and concludes with three case-studies of corporations and social responsibility. Chapter 11 examines the pharmaceutical industry and the energy sector in terms of an emerging paradigm shift from the irresponsible to the responsible corporation, one that accepts the shift from the Global Casino model to the Global Community model.

10 Business Ethics: National and International Dimensions of Social Responsibility

Reality Field 10
Corporate ethics have changed in response to environmental issues and are beginning to evolve as the challenge of global social responsibility enters a new phase of development.

INTRODUCTION

Corporate ethics did not, like Medusa, spring fully-grown from the head of Zeus, but rather evolved from general philosophical and theological inquiries over several centuries of human thought, Eastern and Western, Northern and Southern. As individual societies developed economic, social and religious systems, a sense of right and wrong emerged within each of those societies, expressed through a multiplicity of approaches and value perspectives. That multiplicity is largely, although not entirely, explained by contextual origins, time lags and isolated development of ideas. It may also be explained by fundamental disagreement over value priorities. Suggesting that a fundamental disagreement over the definition of what is right and wrong, however, raises the central question of cultural relativity and whether or not a genuinely universal or global code of ethics is possible.

This chapter presents the argument that universal norms and guidelines are not only possible, but that they have already been developed and are being encouraged, not only by intergovernmental organisations and non-governmental organisations, but also by those multinational corporations which have responded positively to the challenge of social and environmental responsibility. As these corporations benchmark best practices, publish the results and participate in dialogue and policy formulation with governments and NGOs, the values which are implicit in 'best practice' are fostered and dispersed. In so far as these norms are developed in democratic international organisations and culturally diverse groups, they cannot be dismissed as 'cultural imperialism' or the imposition of Western values on non-Western societies.

It is precisely this gradual process of defining *agreed* norms which led to the concept of human rights in its broadest definition and which is spreading from its many points of origin to a global context. For example, when corporations, whether located in Britain, Japan or Brazil, recognised that the demand for improved working conditions had intrinsic merit, they were accepting a value standard which could be universalised. The problem lies not in the value itself, but in implementing procedures, both national and international, which guarantee compliance with stated objectives and which reflect the will of corporations to conduct their affairs in a responsible way. As Sir Geoffrey Chandler, a former Royal Dutch Shell executive and Chair of Amnesty International UK Business, puts it, 'The choice is between the exercise of corporate leadership in developing appropriate company policies, or being forced by public opinion to bring corporate practice into line with the values of society.'[1]

STUDYING BUSINESS ETHICS: THE NEED FOR CONCEPTUAL COHERENCE

In order to discuss business ethics, it is important to understand the variety of sub-topics which are reflected in different approaches to the subject. The study of business ethics is a relatively new sub-discipline in business studies and still lacks the established coherence of older traditions. To begin with, there are three dimensions which must be appreciated in the literature on business ethics: the *international dimension*, in which the research and analytical focus is on multinational corporations with branch operations in many countries, the *national dimension* which targets businesses operating within a single country, and the *meta-ethical dimension* which is composed of applying the writings of classical and modern moral philosophy, theology, economic theory and political theory to the question of determining what is ethical in business.

The international dimension: multinational corporations

The first dimension, the study of multinationals, has been directed almost exclusively at multinationals based in developed countries, primarily Western Europe and North America. They usually represent an attempt to define what is ethical in a cross-cultural and multicultural context, but with a concentration almost exclusively on a rich-nation-MNC/poor-country axis. Research on multinational firms based in *developing* countries and operating abroad is an extremely limited area which needs expansion.

The national dimension: internal governance and external relations

Within the second dimension, another set of distinctions must be made between studies which have focused on the *internal governance* of firms (wages, health and safety regulations, working conditions, personnel development, employment and dismissal policies, assessment and promotion procedures, codes of conduct, the conscience of individual managers) and those which concentrate on *external relations* (quality and reliability of the product, protection of shareholders, customer satisfaction, product guarantees, compensation for inferior merchandise, compliance with national laws and regulations). More recently, a body of related literature has emerged which broadens the traditional concept of external relations to encompass corporate relations with society as a whole (environmental and social issues, protection of stakeholders, the actions of supply chains). Extension of business ethics to include environmental and broader social issues reflects a significant expansion of the concept of business responsibility and it is one which is still unfolding. Very recently the Prince of Wales Business Leaders Forum in the UK has attempted to define a socially responsible business as one in which a company contributes to sustainable development through its core business activities, its social investment or philanthropic activities, and through public policy dialogue. The fundamental assumption of this group and many others, including NGOs and national governments, is that business must go beyond the minimum expectations of operating within the law and adopt a social investment policy that recognises its unique position to foster local community and social development through partnership arrangements.

Meta-ethical dimension: philosophy, theory and common sense

A third distinction may be found in the extensive literature on the religious and philosophical foundations of business ethics, or *meta-ethical theory*. Largely academic in nature, these writings draw upon the rich, historical tradition of logical reasoning in religious and moral philosophy as well as in political, social and economic theory (e.g., the definition of justice, theory of rights, contract theory, structural economics, definitions of morality, values, and ethics, the nature of the state).[2] Any detailed discussion of the philosophy of ethics goes well beyond both the purpose and the focus of this book, but it important to understand that the questions of what is ethical and what is not ethical, what is moral and what is not moral, facing the directors and managers of corporations have been originally framed in the abstract reasoning of philosophers as they attempted to grapple with the complexity of what constitutes an ethical, or a moral, life. This complexity was profoundly deepened as modern industrial society developed and industrial abuses, so vividly depicted in Dickens' description of industrial practice in Victorian England, occurred.

Restraints placed on those abusive practices, both by governments and by corporations interested in reform, were expressed in the form of rights, duties and responsibilities, as in the right to healthy and safe working conditions, the rejection of child labour and the responsibility of corporations for the protection of its working force. The arguments most readily accepted by corporations were pragmatic ones – for example, that companies would be more profitable and more efficient with a secure, healthy, and satisfied adult labour force than they would be with the opposite.

Rather than emphasising inherent rights of workers or a contract between employee and employer, business characterised the reforms in terms of pragmatic 'common sense'. Most business and government leaders who encouraged these reforms were probably not aware that they represented the distinction between a Kantian, rights-based approach to ethics, a deontological emphasis on duty, a Rawlsian theory of justice or a contractarian approach based in Lockean and Hobbesian theory, on the one hand, and a utilitarian, Benthamite, Jamesian approach on the other, but it didn't really matter. The 'common sense' approach has wide popular appeal and is an effective channel for the distillation of ideas which originally flow from the prolific minds of philosophers and theologians.

Tactics, developed in the long struggle to establish a balance between rights and responsibilities in the workplace, often focused on transparency. Corporations which adopted open procedures were celebrated as models and those which engaged in abusive practices were exposed to the critical area of public disclosure and condemnation. Labour unions, civic groups dedicated to reform, government regulations developed through both legislative and judicial processes and corporate self-regulation all played a role in the emergence of widely accepted norms for the treatment of employees.

Over the last century those norms have been gradually extended to include children, ethnic minorities and women. Today it would be very difficult, if not impossible, to find a modern corporation which does not accept these abstract rights, at least in principle, whether or not the Board of Directors has ever read philosophy. What philosophy has so vividly demonstrated, however, is that the search for justice and for ethical norms is a never- ending process. In the post-industrial and post-modern world philosophers continue to expand the notion of rights and the concept of justice in ways which are intended to find a rationale for social action against the mistreatment of living creatures, especially when those creatures are poor, weak and helpless. It also directs this inquiry to the larger context or environment in which living creatures survive, thus also providing a basis for the 'common sense' argument to ensure environmental responsibility and an awareness of the fragility of nature.

Table 10.1 is a summary of the three dimensions – national, international and meta-ethical – to the study of business ethics discussed above and is intended, not to provide easy answers to difficult questions, but to illuminate the complexity of the subject.

Table 10.1 Variable approaches to the study of business ethics

National corporate practice					
Governance	Wages	Health & safety	Employment & dismissal	Assessment & promotion	Working conditions
Individual managers	Codes of conduct	Penalties	Procedures for dissent	Individual conscience	Board of directors
Customers	Product quality	Product guarantee	Consumer protection regulations	Product liability	
External relations	Government	Public			
	Legal compliance	Environmental responsibility Social responsibility			

Table 10.1 Contd.

Multinational corporate practice					
Governance	Wages, work conditions: home or abroad?	Health, Hiring, Dismissal: home or abroad?	Standards in subsidiaries? Standards in suppliers?	Local control, local responsibility?	Liability: home-based or abroad?
Individual managers	Transnational code of conduct	Home-based managers	Board of directors	Local managers	
Customers	Product quality	Product guarantee	Whose consumer protection regulations?	Product liability	
External relations	Home government legal compliance	Host government legal compliance	Home society: environmental & social responsibility	Host society: environmental & social responsibility	
Meta-ethical queries: religious and moral philosophy, economic theory	Nature of justice	Theory of rights	Social contract theory	Defining morality	Defining values

Sources: See note 2.

METHODOLOGIES FOR INVESTIGATING BUSINESS ETHICS

Two fundamental methodologies may be identified in these three approaches, each with limitations. Case studies and descriptions of abuse dominate the procedural, inductive or empirical method of data collection and analysis and they form the primary means by which individual corporations are assessed. The strengths of this approach are well known, but the limitations less so. Too many case studies suffer from incomplete gathering of facts and events as well as distortion and bias caused by preconceived notions of blame and unethical conduct. Far too often, the essential query of an ethical approach – Who has been harmed and by whom? – is restricted to a single group in a scenario which nearly always involves several different groups. Even more problematic are case studies which rely on a single source. In one well-known casebook on European ethics, for example, the case study of one corporation consists of two paragraphs of quotations from a company report.

Logical reasoning is the second approach to 'methodology' in the study of business ethics and is characteristic of meta-ethical inquiry. Unfortunately it is rarely combined effectively with detailed case studies and sometimes suffers from over-reliance on counterfactual conditionals. Existing lacunae between theory and practice is in part attributable to this methodological bifurcation. It also, of course, reflects the gap between ethical rhetoric and practice, or at the very least the gap between aspirations and achievement so often visible in governments and corporations.

Systematic, data-based approaches to the study of business ethics is a third possible methodological approach, but it has been used infrequently and tends to be expressed in the form of surveys and questionnaires distributed to company managers and other personnel, sometimes even to business students, usually in Western Europe or North America, to determine individual attitudes.[3]

BUSINESS ETHICS AND SELF-REGULATION

Although most firms make a strong effort to comply with national laws and international codes, there is a marked preference in corporate statements of principle for self-regulation and voluntary compliance. Table 10.2 overleaf illustrates several possible measures which can be taken by firms themselves, independent of any regulatory body, to support an ethical stance, or what Kevin Jackson calls 'nurturing a cosmopolitan culture of integrity'.[4]

Self-regulation is usually accompanied by a high degree of suspicion from outsiders that firms emphasise self-regulation simply as a smokescreen behind which they avoid legal complications and costly enforcement. Indeed that can be the case, but there is also some legitimate concern by firms over the loss of flexibility which can occur in overly-regulated approaches. Another concern, developing particularly in the health and safety sector, is that of external regulation which is based on faulty or inadequate scientific research (climate change, genetically modified food). Because the companies which have the most to lose, react in the strongest terms to proposed government regulation, their response is seen as a self-interested attempt to block regulation. This action–reaction syndrome,

Table 10.2 Categories for measuring ethical content

Statement of corporate principles	Emergency hotlines
Codes of conduct	Ombudspersons
Ethics training and re-training	Ethics officers
Ethics audit	Ethics committees
Restrictions on personnel data	Adherence to standard established by professional societies
Cultural sensitivity training for foreign operations	Ethnic and gender sensitivity training for employees
Clear supervision and lines of responsibility for branch operations and supply chains	Non-punitive channel of communication for whistle-blowers
Annual reports with a separate ethics report	External audit of contributions to environmental protection and social projects

Source: See note 4.

accompanied by over-simplistic press reports of complicated and contradictory scientific research, tends to create a hostile and adversarial atmosphere not only between corporations and non-governmental groups, but also between governments and corporations. On the positive side, it is usually in this atmosphere of charge and countercharge that more responsible and more knowledgeable voices emerge, and the debate begins to take on the nature of public education on very complex issues. A mediated approach, such as the appointment of a fact-finding commission, in which the interests of a wide range of stakeholders can be taken into consideration often follows. The debate over genetically modified food in the UK in 1999 is a case in point, although by August 1999, the debate had deteriorated to an angry and destructive confrontation between farmers and Greenpeace.

THE GLOBAL DIMENSION: MNCs

Since the late 1960s, multinational corporations have been at the centre of interest in business ethics. These concerns had their origins in radical activism in the United States directed at American multinationals as well as in the reaction of other countries, especially developing countries, to the dilemma posed by external penetration of their economies – loss of national control on the one hand and their desire to attract foreign investment as a development strategy on the other.[5] As the practices of multinationals became more salient, efforts were made to adopt codes of conduct and statements of principle, both intra-firm and internationally, to serve as guidelines for action.

At the same time, national legislation emerged from a wide variety of sources which extended local law, complete with a system of enforcement, to multinationals operating abroad. Neither the international codes nor national law has been able to prevent all abuses in all places, and in many instances corporate reform has been scandal-driven.

Nevertheless, over the past thirty years, a regime of compliance has been constructed which encompasses existing national law which is binding and enforceable with voluntary and non-binding international codes and corporate self-regulation. Reinforcing the trend toward greater responsibility, the multinational corporations have been in the process of re-inventing themselves and behaving in ways which contradict the assumptions of both their critics and their supporters.

The global picture: voluntary compliance codes

International codes of conduct have been evolving since the early 1970s and have recently begun to proliferate. There are now voluntary codes or standards in the European Union, the World Trade Organisation, the Asian-Pacific Economic Co-operation Forum, the North Atlantic Free Trade Association, the Organisation for Economic Co-operation and Development (OECD) and the UN Economic and Social Council (ECOSOC). These codes identify universal, uniform standards relating to apartheid, conflicts of interest, deception, discrimination, embezzlement, fraud, forgery, genocide, slavery, theft and corruption. More recently, particularly in the wake of Nick Leeson and the UK Baring Bank scandal, the growth of global financial transactions, insider trading, executive remuneration, and misuse of pension funds, have been added to the list, especially in the OECD and the IMF. The Caux Round Table based in Switzerland in 1993 adopted an international code for multinational firms in Europe, North America and Japan which identifies five basic principles that go well beyond the earlier codes which focused on more restricted abuses. Those principles are stakeholder responsibility, social justice, mutual support, environmental concern and avoidance of illicit operations and corrupt practices.

Individual corporate codes of ethics or codes of practice tend to be a mixture of technical, prudential and moral imperatives expressed in a general statement of principles and with varying degrees of enforcement. Critics of these codes have pointed to their avoidance of specific issues, the failure to include enforcement sanctions, their overly-generalised content and their avoidance of certain issues, such as profit maximisation.[6] However, that trend may be changing as an increasing number of corporations identify 'profitability' and 'shareholder interests' in the mission statement as central goals, but also accompanied by statements of corporate responsibility for their employees, the environment and the community.

The enforcement dilemma

Whilst it may well be the case that rules for the operation of corporate entities should be the same for both national and multinational corporations, the issues are different with respect to the feasibility of enforcement. In the case of national companies, national laws on bribery, false advertising, malfeasance, mistreatment of employees and other forms of unethical practice serve as the mechanism by which national companies are held accountable.

In the case of companies located in North America, Japan and Western Europe, legal systems of enforcement and restraint operate within the context of abundant state resources, high levels of education, a tradition of voluntary reform groups, an unfettered press, public officials who can be removed from office through the electoral and/or litigation processes and a relatively efficient structure for detection, trial and punishment for abrogation of the law. In countries where these systems and structures are not yet in place or where they are severely constrained by oligarchic or authoritarian political traditions, by underdeveloped legal structures, by low levels of public education and by ineffective or corrupt enforcement procedures, national and local businesses corporations can function without regard to formal and legal codes of conduct. In many instances in both developed and developing countries, national codes and laws exist on paper, but because they are not enforced on a regular and impartial basis, they become an avenue for discrimination and abuse. The Simon Jones case in Britain provides a clear example of the problem, even in countries where protection of workers is highly developed.

CASE: SIMON JONES: THE DANGERS OF CASUAL WORK

Simon Jones, a 24-year-old intermittent student who was receiving social benefits for unemployment, was placed by a Brighton employment agency in a temporary job.[7] He was employed by Euromin, a Dutch-owned firm, to unload cobblestones from a Polish ship named *Cambrook* which was anchored at the Shoreham docks in the UK. On 24 April 1998, Simon was working in the hold of the ship, hooking bags of cobblestones to chains that were attached inside a crane's open grab. Although it was not necessary to use a grab in this operation, its use saved time and expense for the company. The grab and chains were lowered too close to the hold and the grab closed on Mr Jones's head; within a few hours, he was dead.

The police arrested the general manager of Euromin, James Martell, and the crane driver who were both subsequently released without charge. The Health and Safety Executive (HSE) issued two prohibition notices and an improvement notice against Euromin, prohibiting the use of chains inside the grab and operating the crane in areas where others are at risk and requiring it to introduce training and supervision. Euromin, the private owner of the dock, was in violation of health and safety regulations in the UK and in Holland. Personnel Selection asserted that it was misinformed regarding the nature of the work and the Department of Trade and Industry decided that there was insufficient evidence to take action against Personnel Selection, the agency which placed Simon Jones in the job. In September, demonstrators occupied Shoreham docks, scaling the lighting rigs to hang banners protesting the death and the lack of protection for casual labourers and urging tougher penalties for industrial injuries. A Simon Jones Memorial Campaign was launched to pursue these reforms.

In January 1999 the Crown Prosecution Service decided not to charge Euromin or its general manager with manslaughter because of insufficient evidence. A full review of the case has now been ordered by the Industry Minister and Mr Jones's family will seek a judicial review of the case if the Crown Prosecution Service fails to act. According to the British press, the government is to 'begin consultations soon' to provide much higher fines against companies where criminal negligence by employers causes death at work. At present, fines are at a very low level for breaching health and safety law. The government has also boosted spending on the HSE by 17 per cent in real terms but a severe shortage of inspectors may inhibit effective action.

Although the Simon Jones case has not yet been resolved, moral and ethical questions could certainly be raised about the practices of the Dutch company, the British employment agency and the British government agencies responsible for protective legislation, but the point remains that legal redress is possible, especially in cases which have been made visible by protest and subsequent press coverage. In this case, reform can be pursued in the country where the abuse has occurred. If that had not been the case, then reform could have been pursued in Holland where Euromin is based and where European Union law would dominate. But if Euromin had been unloading its cargo in an African or Asian state, extra-territorial application of Dutch law would become a central issue; hence the tendency to demand an ethically-based response from firms in those instances where the legal route is problematic.

ETHICS AND MULTINATIONALS: WHY THE NATURE OF THE STATE MATTERS

More serious problems are posed by abuses occurring in countries where public protest is prohibited, where the press is controlled and where the legal system and/or the government is corrupt. In these cases, the current tendency is to pursue responsibility for abuse in the home base of multinational corporations. Although this tactic may result in monetary compensation for those directly affected, it does nothing to reform the countries in which such abuse is either not prohibited or where it is prohibited but not enforced. The problems posed by uneven application of law, in turn, creates a situation which is fraught for multinational companies, especially if they are bound by established law in developed countries where enforcement is effective and their competitors located in the host country are not.

Because the conduct of multinational corporations based in developed countries has received far more public attention than the conduct of national companies in developing societies or of multinational corporations based in developing countries, developed-country MNCs have been subjected to much greater surveillance. Stringent national legislation as well as international codes of behaviour fostered by international organisations and by the corporations themselves have led to a situation in which large multinational corporations, particularly if they are based in the USA, become the target of global watchdog activities. Because their conduct is relatively transparent, i.e. compared with national companies, some large multinationals have led the way in the development of ethical best practice, a trend which is both desirable and understandable.

Greater concern, however, must lie with those companies which remain hidden from the headlines. They pose a serious threat to the development of human-rights-based codes of conduct and to the competitiveness of foreign-based companies. Yet they have received very little attention, precisely because their operations and practices can be removed from the glare of publicity and from the application of law.

Human rights groups, environmental groups and labour unions, frustrated by their inability to influence national governments which do not monitor the conduct of businesses for compliance with the law, focus their attention on Western-based corporations in the hope that inadequate local laws and lack of enforcement can be by-passed. By forcing multinational, Western-based companies to bring their foreign-based operations into conformity with the law at home, advocate groups hope to force a change in national practice in the host country or to achieve an extra-territorial application of the laws in country A to the practices in country B. However, what has been the case more often, is that MNCs, particularly successful ones, are held to a standard of performance which is not applied to their host-country competitors. Large, successful and wealthy MNCs become an easy target for threats of litigation and the payment of compensation which is far above what constitutes current practice in the local economy. And they also become a conduit for extra-territorial application of national law, especially when the labour standards of one country are applied through the multinational to supply chains which are based entirely in another national economy, sometimes not even that of the host country. The case of Tommy Hilfiger will serve to illustrate the point.

CASE: TOMMY HILFIGER *ET AL.*: WHOSE RESPONSIBILITY FOR WHAT?

Outline of the case

Tommy Hilfiger is an American clothes designer who subcontracts the sewing of his clothing to Asian companies operating garment factories in the Pacific Northern Mariana Islands. The practice is widespread and followed by 17 other American companies, including Gap, Sears Roebuck, Wal-Mart and Oshkosh B'Gosh (children's clothing), who have been targeted for legal action along with Tommy Hilfiger. American human rights groups such as Global Exchange, Sweatshop Watch and the US Union of Needle-trades, Industrial and Textile Employees have brought legal action in the United States against these US retailers who are alleged to be responsible for the abuse of workers employed by Asian sub-contractors. Workers from China, the Philippines, Bangladesh and Thailand are recruited and hired by Asian-based companies which have set up sweatshops in the islands.[8]

Asian-based companies, as well as American companies, are attracted to the Marianas, in particular to Saipan, because the post-Second World War commonwealth arrangements with the US provided that island exports would be exempt from US duties and quotas. Although the companies which are alleged to have mistreated workers are Asian and the local government is responsible for labour laws, the suit against American retailers has been brought in the US courts where an attempt will be made to apply the American minimum wage, which of course is much higher than wages in China, Bangladesh, Thailand, the Philippines or the Marianas.

The suit also seeks $1 billion (£610 million) on behalf of the 'guest workers', those workers who have paid recruiters in Asia to secure jobs, to finance their transportation to the islands and to provide their accommodation once employed. Although the case has not yet been adjudicated, it provides an interesting case study for illuminating the complexity of ethical issues in the international dimension.

Background: political and legal status of the Commonwealth of the Northern Mariana Islands

The Marianas are an archipelago of 16 islands with a population of approximately 20,000 people who reside on six islands, including Saipan, the administrative centre. The Marianas, forming a part of Micronesia, were held as colonies by Germany before the First World War. With the defeat of Germany in 1918, they were administered (with the exception of Guam) by Japan under a League of Nations mandate during the inter-war period. With the defeat of Japan at the end of the Second World War, they were transferred by the UN Security Council to a UN trusteeship with the United States as administering authority.

As the demand for self-government throughout Micronesia grew, limited legislative authority was conferred in 1965 on the Congress of Micronesia, consisting of a Senate and a House of Representatives to which the Marianas elected three representatives. In 1972 a Marianas Political Status Commission initiated negotiations with the US that resulted in the 1975 covenant establishing a Commonwealth of the

Northern Mariana Islands in political union with the United States, similar to the arrangements of the British Commonwealth possessions, in which local self-governance, including foreign policy, was established, but responsibility for defence was retained by the US. As part of its general economic development support, the US also agreed that goods produced in the Marianas would enter the US market duty-free.

The Commonwealth's first governor and legislature assumed office in January 1978. Negotiations were conducted in 1979–80 to complete the 'compacts' of association for three additional Pacific Trust units (the Marshall Islands, the Republic of Belau and the Federated States of Micronesia) and which were approved by UN-monitored plebiscites in 1983. All four compacts were then approved by the US Congress in 1985. In June 1986 the UN Trusteeship Council endorsed by a 3–1 vote the position that the US had satisfactorily discharged its obligations and that it was appropriate to terminate the trusteeship. The Soviet Union objected to the defence arrangements and to US economic development policy and threatened to use its veto in the UN Security Council to prevent the US trusteeship responsibility from being discharged. The US, with the support of the Commonwealth of the Northern Marianas, the Federated States of Micronesia and the Marshall Islands, then declared the compacts for those three states to be in effect and in 1986 the Commonwealth of the Northern Mariana Islands ceased to be a component of the Trust Territory, with its residents becoming US citizens. After the collapse of the Soviet Union, the Russian position with regard to trusteeships was abandoned.

Loss of jobs: the issue in the Marianas and the United States

Very often, multinational corporations have defended the continuation of practices which do not comply with home-based policies in the name of protecting workers' jobs in the host country. In this case, the loss of jobs is relevant to both countries and adds a further layer of complexity to the case study. In a developing economy such as the Northern Marianas, the loss of both jobs and exports would have a significant and negative impact on development. Furthermore, although the wages paid in the Marianas by the Asian-owned companies are lower than in the US and in the Marianas for its citizens, they are higher than the Asian workers would receive in their respective countries of origin. Hence the sub-text for this case is not just a foreign-based multinational operating in a host country, but workers of third-country origin (e.g. China) travelling to another country (the Marianas) in seek of higher earnings by working for a home-based company (the Asian recruiting companies) located abroad and contracting piece work from a multinational based in yet another country (US).

In the United States, the clothing industry, with very few exceptions, no longer maintains a significant manufacturing base in the US. In order to remain competitive, these firms have reduced jobs in the US and moved offshore, scaled back their operations or never owned manufacturing facilities. In 1999, just a few days after the suit was brought against Tommy Hilfiger *et al.*, Levi Strauss, another large clothing manufacturer, cut 30 per cent of its workforce and announced that it would shift production outside the US so it can concentrate on product design and marketing.[9] Four plants will be closed in Texas, two in Tennessee and one each in Georgia, Arkansas, Virginia, North Carolina and Ontario, Canada.

Other large firms such as Russell Corp., Cone Mills Corp. and Fruit of the Loom Inc. are making similar reductions. Employees in these firms tend to be unskilled or semi-skilled, many are recent immigrants to the US and the firms are often located in states with weak state economies such as Georgia, Arkansas, North Carolina and Tennessee which have provided incentives to corporations willing to create employment in depressed areas.

The Union of Needle Trades represents many of the garment industry workers in the US who are facing job reductions on a large scale. In a very real sense, the gain of jobs in the Northern Marianas is a loss of jobs in the United States, just as it was a decade earlier when US garment companies began to relocate in Mexico. The action of the union must be seen as a reflection of this problem as well as its concern for the treatment of Asian workers in the Northern Marianas.

What is the ethical and legal dilemma?

If the abusive treatment of women working in sweatshops were occurring in a factory owned and operated by any of the American companies and occurring in a country whose citizens are also American citizens, the responsibility would be clear. The American companies could be required, under US law, to apply American standards for working conditions. Because the companies which have been accused of abusive practices are Asian companies, have employed non-US citizens and have located in the islands with the permission of the local government which has responsibility for economic development as well as for working conditions, the question of liability becomes more complicated. The Asian-owned companies are also contracting work from the American retailers; hence the issue centres on the operation of supply chains which are neither owned nor controlled by the retailers and which are not based in the Commonwealth.

Moralists would cut this Gordian knot by asserting a theory of universal rights (Kant) or a theory of universal principles of justice (Rawls) to be applied to the workers, regardless of their national origins, and would correctly assert that the Asian companies, the governments of the Commonwealth and the United States and the American retailers have failed to protect these rights. Deontologists would assert the notion of duty and argue that some acts are morally obligatory regardless of consequences in practice and therefore the US retailers should force the Asian recruiters to abandon their immoral practices or lose their contracts. Utilitarians would emphasise that what is moral is what serves the greatest good for the greatest number of people; therefore it is moral to improve the working conditions. Contractarians would argue that a social contract exists between the people and their government which implicitly requires their protection in exchange for their acceptance of authority, but of course the Asian workers are neither US citizens nor citizens of the Commonwealth of the Northern Marianas.

In each of these instances of abstract, principled reasoning, the conclusion would be that it is immoral and unjust to mistreat workers. The difficulty lies in trying to determine who has legal responsibility for the failure to protect and who should pay for increased wages, improved working conditions and more transparent methods of recruiting foreign workers. The common and widespread assumption is that 'wealthy and powerful' corporations alone should bear the responsibility. It is satisfying to make that assumption and it is morally and politically correct in the United States and Western Europe to do so. But is it just? Is it equitable? Is it moral? For whom? And finally, is it the only assumption that should prevail?

Ten difficult questions: who has harmed whom?

1. Should Asian companies, operating as subcontractors to American retailers in a third country be subject to American domestic law or

to the domestic laws of the third country, e.g. China, Bangladesh, the Philippines or Thailand? To all?

2. Should US domestic law, including its minimum wage, be applied to the Marianas, despite the latter's self-governing status, its economic need and its lack of human and natural resources? Should the US government withdraw the duty-free status of exports from the Marianas to the US market to remove the incentive for US companies to operate there?
3. Should the US retailers end an unethical practice by cancelling their subcontracts with the Asian suppliers, even if island jobs and exports are lost and the Asian companies then withdraw? Should they do so to end the possibility of litigation?
4. Should the Commonwealth government adopt national legislation and enforcement mechanisms to protect both national and foreign-born workers from abusive practices?
5. Should the governments of China, Bangladesh, the Philippines and Thailand adopt similar legislation to control the practice of their national companies operating abroad?
6. Are the human rights groups involved serving justice and behaving morally by bringing legal suits against the retailers, but failing to do so with respect to the government of the islands and the Asian subcontractors?
7. Was the motivation for bringing suit in the United States driven by trust in the impartiality of the American legal system and the willingness of American companies to comply with the law or by its reputation for ordering very large sums of money in damage cases?
8. Is the decision of Global Exchange, Sweatshop Watch and the Union of Needle-trades to seek compensation only for foreign workers equitable? Should the request for compensation have included the Commonwealth government for any loss of exports which ensue? Is the opposition of the Union of Needle-trades to the export of American unskilled and semi-skilled jobs in the clothing industry in favour of jobs in lower-wage areas ethically relevant?
9. Is the *Guardian* newspaper journalist, who reported the suit in the UK and named five out of 17 companies charged and also failed to name the Asian companies who are alleged to have exploited workers, behaving in a socially responsible manner? Should he have named all of the companies? What is the ethical significance of a 'naming and shaming' strategy if it is applied selectively?

10. Assuming that the US retailers are judged to be responsible for the actions of their suppliers and ordered to pay compensation to the foreign workers, is justice served by allowing the Asian companies who have previously extracted fees from the workers, to avoid retribution? Should the Asian companies be required to contribute to retribution costs from the profits made on recruiting foreign workers? Does the government of the Northern Marianas bear any ethical responsibility for its 'guest' workers?

In order to answer these questions, more information on the details of the case would be necessary but what the case highlights is not only the complexity of the issues involved, but also the pitfalls encountered in legalistic and punitive approaches.

Would a mediated conflict-resolution strategy produce more desirable results? Assuming a first and universal principle of ethics, i.e. that workers should not be exploited and mistreated, would alleviation of their suffering have occurred more rapidly if Global Watch and Sweatshop Watch had attempted a solution based on multiple rights and multiple responsibilities? The overwhelming advantage of a mediated and equitable solution is that all parties must be represented. Perhaps with the assistance of the Commonwealth government and disinterested third parties, such as other NGOs and intergovernmental organisations, a negotiated and agreed compromise could be devised which would better serve the interests of all parties. What is desirable is to end the abuse by means which would preserve the incentives for the US retailers to continue export production in the Northern Marianas, thus protecting both the local economy and the foreign workers as well as improving the practice of the Asian recruiting firms. It would not, however, protect American garment workers from job losses, but it could possibly result in the creation of new jobs, probably not unskilled and semi-skilled. By casting the solution in materialistic and legalistic terms, the opportunity for significant and lasting reform may be lost.

THE SOCIAL DIMENSION

Most of the early writings on business ethics were concerned with internal corporate governance issues, the responses to which developed over most of the nineteenth and twentieth centuries as the rights and responsibilities of corporations and workers were negotiated. By the end of the twentieth century these rights and responsibilities were written and codified in law and have universal acceptance and recognition in the UN Universal Declaration of Human Rights, adopted by the UN General Assembly in 1948. Twenty-eight years elapsed before the key elements of this non-binding resolution were embodied in binding international agreements.[10] The two human rights covenants which comprise the Declaration, one embodying economic, social and cultural rights and the second, embodying political and civil rights, were brought into force in 1976. Those rights, as abstract principles, are widely supported by the member states of the United Nations and by human rights' organisations, but in practice, they are often selectively applied, ignored or breached in many, if not most, countries.

Because they embody such a wide range of individual rights and because they reflect social, economic and democratic ideals associated with Western states, they have at various times since 1948 been treated with great suspicion. They are declared to be universal, but the distinction between theory and practice is of utmost importance and they must be seen as ideal goals and values or a standard of excellence, rather than strict practice. Indeed, it is only as an ideal goal that advanced countries could claim recognition of the 'right to protection against unemployment or the right to equal pay for equal work' (art. 23). Nor is it persuasive that all reform groups would share the description of the family as the 'natural and fundamental group unit of society' (art. 16). Islamic fundamentalist and Hindu states would have difficulty with the notion that 'marriage shall be entered into only with the free and full consent of the intending spouses' (art. 16). The European Union's employment guidelines would violate the prohibition against discrimination on the basis of national origin (art. 2).

With respect to the relationship between 'human-rights-based ethics' and corporations, especially multinational corporations, it will evolve in the same direction and with the same applicability as it has for governments. That is, the gap between theory and practice, between ideals and practice will not only continue, but will be tolerated comfortably, except in those cases where the abuse is egregious or where it clearly violates existing national and international law. Having said that, responsible corporations have extended the range of what is considered ethical, albeit within the boundaries of what they view as the most important functions of corporations. The following discussion of Shell identifies those priorities and contradictions for one firm.

CASE: CORPORATIONS AND SOCIAL RESPONSIBILITY: SHELL UK LTD

> The public demands from us the highest standards of ethical and environmental responsibility. They expect us to take a long-term interest in the economic and social well-being of the wider community, including the international community, and reflect this in sensitive development of the world's natural resources.[11]

> Shell UK has a long-standing commitment to corporate social responsibility and to upholding high ethical standards.[12]

> We do not seek to position Shell UK as an all-embracing corporate 'nanny' stepping into areas where government, local authorities and others cannot or will not tread.[13]

Three quotations from the 1998 *Report to Society* distributed by Shell UK Ltd serve to highlight the dilemma implicit in any discussion of ethics and social responsibility in business, that of establishing the boundaries of responsibility.

The question raised here has to do with the limitations of corporate responsibility. Shell UK is Britain's largest oil, gas and petrochemical company, its fifth largest newsagent, one of its 'top four sandwich retailers' and it employs 6,700 employees directly and 19,000 indirectly. It produces 19 per cent of Britain's crude oil, 17 per cent of its gas, 18 per cent of its oil products and 15 per cent of its petrochemicals. Annual sales for the Royal Dutch Shell's group in 1990–1 made it the 28th largest 'economic unit' in the world, ranking well above Argentina, Saudia Arabia, Thailand, New Zealand, Greece and Ireland.[14] As the confrontation between Greenpeace and Shell described in Chapter 5 illustrated, Shell has become an increasingly responsible company, implementing many policies to improve its environmental record and to contribute to the local areas in which it operates.

In the first edition of its *Report to Society*, the firm has attempted to place its contributions in a very broad perspective, one not often cited by its critics. For example, in 1997 it paid £380 million in direct taxes and £3.8 billion in royalties, VAT and other indirect taxes as well as contributing to the local economy of Aberdeen through its 700 local suppliers. Its health, safety and environmental data are verified by external auditors and published; it has formed small management teams responsible for benchmarking the best practices in major operations such as flaring and venting, environmental certification, contractors' contributions, energy efficiency, solid waste reduction, drilling discharges, reducing drill cuttings on the seabed and reducing emissions. In fact, Shell has reduced its CO_2 emissions by 22 per cent since 1991 and it is careful to point out that the oil and gas industry represent only 2–4 per cent of total greenhouse gas emissions from the UK.[15]

With respect to its commitment articulated in the Statement of General Business Principles 'to take a constructive interest in broader social issues', Shell provides information on its relationship with its employees: personnel development, training programmes in Aberdeen for Nigerians employed in its liquefied gas operation in Nigeria, percentages of ethnic minorities and women employed (2 per cent and 13 per cent respectively), harassment and grievance procedures, drugs and alcohol programmes, subsidised lunches, counselling services, flexible working, pensions above the Retail Price Index, and the first woman appointed to the Board of Directors in 1997.

The *Report to Society* also details community activities such as tree planting in Manchester, land transfers to extend the Red Rose forest, sponsoring Local Agenda 21 conferences and networking, support for projects through the Better Britain Campaign, a youth work experience programme with the Department of Training and Industry and Enterprise Councils and grants to rural community groups in Scotland with a population of less than 2,000. Shell UK, of course, is not unique in this list of corporate activities, nor is it the best, but it can serve as a model of a firm which is reforming its practices and which represents the trend not only toward increased responsibility, but recognition of the need to provide accurate information to the public. Although its critics dismiss these efforts as 'nothing more than a public relations campaign', that is clearly not the case, especially on environmental issues.

A statement of general business principles, which applies to the Royal Dutch/Shell Group of Companies, identifies seven areas of commitment: *profitability* (defined as a measure of efficiency and value that customers place on Shell products and services; investment which takes into account social and environmental considerations); *integrity* (prohibit bribes, avoid conflicts of interest, accurate and fair business transactions and accounting); *politics* (obey local laws, prohibition of payment to political parties, organisations or their representatives and participation in party politics); *health, safety and the environment* (sustainable development and a 'systematic approach to health, safety and environmental management to promote continuous improvement'); *community* (a constructive interest in broader social issues not directly related to the core business); *competition* (support for free enterprise, 'seek to compete fairly and ethically within the framework of applicable competition laws') and *communication* ('open communication and dialogue with the outside world'). Shell has clearly identified four general goals for the group of companies as well as their core values. *Table* 10.3 presents a summary.

Table 10.3 Royal/Dutch Shell Group of Companies: goals, values, principles

Core values	*Goals*	*Principles*
Honesty	Sustainable development	Profitability
Integrity	Uphold highest consumer standards	Integrity
Respect for people	Protect working conditions	Politics
	Promote best practice among contractors, suppliers, co-venturers and partners	Health, Safety and the Environment
		Community
		Competition
		Communication

Source: adapted from the text of *Report to Society*, Shell UK Ltd.

Caveats and limitations

While these general principles are desirable from an ethical point of view, they are accompanied by a very clear statement of Shell's opposition to any proposals, including those of the European Parliament, that would increase its costs through the tightening of standards, especially air quality standards. Although Shell is working on the development of British standards, the implication of its opposition to extra costs is clear. Implicitly what Shell is saying, and which is similar to many other firms, is that there is a limit to the expenditures which can be made and that Shell will decide where those limits are. As the third quotation at the beginning of this discussion demonstrates, Shell's commitment to issues of social responsibility is limited and the firm has done less in this dimension than in its response to environmental challenges. The *Report to Society*, for example, cites an annual MORI poll of corporate social responsibility in which looking after employees is number one and becoming involved in local events is fifth out of six response categories.

Shell's position on the relative importance of social issues relates to a much wider debate over the parameters of corporate responsibility in general, with positions that range from a belief that corporations have responsibility only to make a profit for their shareholders, to a belief that corporations must apply a very broad range of standards in human rights not only to their own operations but also to all their suppliers. It is difficult to imagine any corporation applying all the rights provided in the Universal Declaration of Human Rights to itself and to all its suppliers, especially when those suppliers may range into thousands of firms and sub-contractors. Indeed, it is difficult to imagine any government so doing. What remains then is to do what responsible corporations are already doing in response to social issues: extend best practices wherever it is feasible; encourage suppliers, contractors, co-venturers and partners to do the same; provide credible evidence that they are complying with their own stated goals; give reasonable support to well-informed groups seeking reform; support initiatives from governments and international organisations which take into account corporate interests as well as others; and recognise a special responsibility, as Shell has done with rural Scottish communities, for the needy.

ETHICAL TRADE INITIATIVES: A FUTURE CHALLENGE FOR CORPORATIONS?

Non-governmental organisations, especially those which are trying to promote development in lesser developed countries, have urged corporations to consider social aspects in trade issues, as well as in production and marketing. National governments have begun funding ethical trading schemes which in some instances are directed toward economic development and in others toward extending 'best practice' to Asia, Africa and Latin America. Fundamentally, it is seen as trade in goods produced under conditions that are socially and/or environmentally as well as economically responsible. In its broadest definition, ethical trade could include human welfare (wages, working conditions), social welfare (protection of small, rural peasant communities, sustainable rural livelihoods) and environmental welfare (animal protection, organic farming, ecological responsibility, biodiversity). In effect, the concept of ethical trade is simply applying the ethical agenda which has been discussed throughout this chapter to trade issues. It may be related to almost any commodity which is traded across national boundaries and is usually attached to some sort of labelling or brand-mark certification scheme in which the producer is given a presumed market advantage with consumers by identifying the product as 'ethically sourced'. There are many varieties of ethical or 'fair trading' schemes, but the market share of these types of goods tends to be very low, 0.01 per cent of global trade.[16] Examples of this approach would be the fair-trade marks of Max Havelaar, Transfair and the Fairtrade Foundation which have been applied to bananas, coffee, cocoa, honey, chocolate, sugar and tea.

Auditing and certification systems are being developed for a range of trading sectors: forestry, organic agriculture, floriculture, food products and some luxury goods. Some ethical trade schemes are commercial and some are community-owned. It is the latter approach, usually in the form of community co-operatives or primary producer groups, which relates most clearly to the developmental objectives and are seen to benefit marginalised people. However, there are many problems associated with using trade as a development tool and its successes are limited.[17] The limitations imposed by government price subsides, production inefficiencies, higher prices to cover the costs of auditing and certification, small-scale production and inattention to marketing do not suggest a viable future for many of these schemes as an avenue to economic development. However, when the idea is applied to encouraging responsible behaviour in conventional trading chains, particularly when adopted by mainstream retailers and importers, its prospects for success are greatly enhanced.

LARGE CORPORATIONS: THE POWER/WEALTH RESPONSIBILITY RATIO

In concluding this chapter, it is important to address its underlying assumption, i.e. that business has a special or unique set of responsibilities for social and environmental protection. An argument for responsibility based on self-interest considerations such as profit, a healthy, well-trained and satisfied workforce, quality products and legal compliance is widely accepted among businesses. Equally unproblematic is the notion of individual ethical behaviour as applied to corporate personnel, managers and directors. Extending that responsibility beyond the immediate and direct interests of business requires another justification, a responsibility ethic which is linked to corporate profits and power. For example, it is customary in foreign policy to link power and responsibility, i.e. the wealthier states in the international system are expected to assist the poorer states through development aid and the most powerful states are expected to protect the weaker ones through alliances and coalitions. The same ratio of power and wealth to responsibility is being applied to corporations: the wealthier and more powerful the corporation, the greater its responsibility. This Power/Wealth Responsibility Ratio may well serve as the moral maxim for corporations in the next century.

11 Globalisation and the Environmentalised Corporation: Present Exemplars and Future Prospects

Reality 12
The market has created in recent decades major transnational enterprises which have become bell-wethers for the development of a new generation of global ethical corporations. Their experience suggests a way ahead for business.

As world trade has expanded, attempts to derail environmental laws at the state level occur. We argue that environmentalisation, and product and plant safety, are inherently beneficial to industrial profits. This fact is evident in many of the advanced technology companies which have become prominent in the last years of the century.

This chapter analyses some of the businesses which together already constitute a growing and successful group of global environmentalised corporations. Based on their experience, it suggests guidelines for establishing a cross-sector standard for ethical operations of business.

INTRODUCTION: THE END OF THE MILLENNIUM AND THE BEGINNING OF SUSTAINABLE GLOBAL BUSINESSES

In 1998, at the time of writing, the multilateral agreement initiative (MAI), a concept stemming from early years of the World Trade Organisation, posed in its draft form a potential threat to national environmental regulatory regimes. In the name of trade liberalisation, WTO's committees proposed to subject all product quality, environmental and safety regulations affecting imports across national boundaries to the test of whether they discriminated against goods produced to different standards in member states. Those found to do so would have to be repealed.

The MAI drew the criticism of, among others, a broad spectrum of environmental NGOs and experts. Fully implemented, it might trigger a comprehensive rollback of the gains since the 1970s in establishing the right of national states to regulate product safety and environmental outputs in their respective territories. Happily by the end of 1998 the OECD committees which had introduced it to WTO had withdrawn the draft, so flawed were its objectives and implementation procedures.

This withdrawal reflects the fact that trade liberalisation itself constituted a process increasingly in doubt in the late 1990s. Recurrent trade and currency crises racked Asian and emerging markets. Doubt grew as to whether free trade among nations, the much vaunted core of late-twentieth-century prosperity, could be sustained. Computer-driven currency-trading systems allowed aggressive manipulation of sovereign states' destinies by short-term profit optimisation systems and their proprietors. Even their most successful exponents, such as the financial speculator George Soros, realised that the procedures involved genuinely brought the global trading systems into chronic instability. The agenda for international regulation and control, curbing the excesses of globalisation, was clearly coming into place.

Prior to this turmoil, ethical transnational sectors had emerged and become established as global self-regulating entities without all-embracing treaties or institutions. The principal such sector, resulting from the phenomenal success of medicines and health-care technology in the post-Second World War period, is the world pharmaceutical industry.

THE FLAGSHIP ENVIRONMENTALISED SECTOR: THE WORLD PHARMACEUTICAL INDUSTRY

The world pharmaceutical industry can be seen to have committed to high environmental performance earlier and more comprehensively than any other global manufacturing sector.[1] Three main reasons for achieving this present themselves in the history of the industry's development. First, the present regime of scientifically tested drugs and medicines is comparatively recent. Until the last half-century, drugs were not subjected to systematic scientific scrutiny and comprehensive bio-engineering. With the discovery of penicillin and other modern antibiotics as well as vaccines, all this changed. For the first time in history the medical profession could rely on a proven procedure for ensuring that their treatments would have the predicted impact on illness.

Second, drugs and medical aids based on biochemical technology must be perceived by consumers and medical systems as 'pure'.[2] Their safety within specified use guidelines must be structured into their development. In all local marketing conditions, consumers and doctors aim to acquire medicines which are manufactured to the same standard. Where there are problems in gaining such assurances, the international press repeatedly focuses on the perceived mismatch between imputed purity and local relaxation. Consequently by the end of the twentieth century, established known methods across the pharmaceutical industry exist for agreeing on a range of product standards.[3]

Third, product purity is assured in a manufacturing setting. The conditions of production are seen as equal in value to the quality of the final product. As a secondary impact of the prime value of purity in production, manufacturing plants have become environmentally managed, regardless of location. In general, by the end of the century, pharmaceutical plants built by multinational enterprises in developing countries operate to the same health, safety, quality and environmental standards as in the home country, if not always at the same price.

To have it otherwise is to court disaster. If, for example, it were reported that a Brazilian pharmaceutical plant owned by American-owned X corporation was producing a known drug to a lower standard or with environmental degradation intolerable in an American setting, it would have major marketing impacts in all its markets. American consumers would become uncertain that the source of their own supply might not be the defective overseas plant. Lack of confidence of professionals and consumers alike would undermine the company's reputation. Even more critically, sources of capital and investment in the information-sensitive stock and bond markets would be priced upward or dry up entirely, as additional risks were factored in to investor's calculations.

Consequently, we can indicate a list of preconditions which such 'high purity' global manufacturing sectors will approximate (Table 11.1) Other sectors which may be approaching these standards include silicon microchip production. Relatively few industries can command such a comprehensive product assurance methodology and reputation as ethical pharmaceuticals.

Table 11.1 Benchmarks for an internationalised environmentalised manufacturing–trading sector: profile based on the case of the global pharmaceuticals sector

Benchmark number	*Explanation*
1. Existence of an established industry association with credibility among all leading producers.	Required for agreeing voluntary standards of performance, e.g., in product development or environmental performance. Often a key interfacing institution with government regulators.
2. Agreement within leading companies within sector that common product and process standards are essential for consumer, government and stakeholder confidence.	Without agreement, the industry remains fragmented. No trust of new products can be predicted. High research and development costs for new products thus become uncommercial.
3. A standardised cross-industry product-development procedure.	New products must be seen not only to perform as specified but also to be the outcome of manufacturing processes known to not impose health costs on producers, communities or patients. These would include environmental costs.
4. Fundamental commitment among all industry members to make consumer interests primary.	This condition is the essence of ethics in business: all accepted producers of these products must be qualified to form this commitment.
5. Government-regulated new-product approvals procedure.	While the industry is deemed to be diligent in its commitment to product effectiveness and safety, this diligence must be comprehensively verified product by product, at all points of marketing release, and throughout the product's life cycle.
6. Officially established product-licensing procedure.	Following market release, the licence must be renewed as evidence about consumer impacts from product use becomes available. The product can be withdrawn from market by either producers or government regulator.
7. Product quality-assurance systems.	
8. Product contamination avoidance strategies.	
9. Post-marketing product surveillance and reporting systems.	
10. Regulated marketing procedures with systematic review of methods of dissemination.	This area is subject to the widest variance of all guidelines, in that developing-country drug availability is often managed without the controls normal in countries with a higher incidence of qualified physicians.
11. Ongoing monitoring of product impacts and secondary effects, including side-effects in consumers.	

It is possible using the best-practice model of the global ethical pharmaceuticals sector to establish benchmark conditions for determining whether a particular industry is environmentally manufactured and/or traded.[4] Primary importance attaches to the role of voluntary associations including the leading competing corporations in the sector.

Only such associations, controlled by their members, can adopt without coercion practices and standards which then percolate through to all new members. Today, all major industrial sectors are represented in both national and international standards negotiations by international trade associations. These have substantial influence over both governmental regulators and member companies. In some cases they can bring about upgrading and commanalties in product standards and licensing procedures. For environmental performance norms, without doubt the industrial associations have been and will continue to be powerful research centres and communicators both to their members and to the public.

If we compare two industrial sectors on these criteria, Table 11.2 overleaf demonstrates that substantial differences separate industries in their environmental performance. However, it also suggests methods by which lagging sectors (energy, transportation) can begin to benchmark themselves against leading environmentalised industrial sectors such as pharmaceuticals.

PERCEPTIONS OF PRODUCT PURITY AND EXCLUSIVITY AS PRICE DETERMINANT IN INTERNATIONAL TRADE

Increasing industrial interest in ethics is occurring along these lines. By 1999 fossil-fuel corporations like Shell and BP were increasingly committed to the environmental agenda. They were amplifying their product portfolios to include renewable energy sources. Between 1993 and 1998 Shell performed a *volte-face* in its publicly communicated corporate environmental profile, going from a position of 'we are already doing it' to one of 'we are comprehensively reviewing with our stakeholders' involvement our community and environmental impacts world-wide and on a continuous basis'.[5] In between had been two public relations and corporate marketing disasters. The failed attempts to dispose of oil production platforms at the bottom of the North Sea and to dismiss as irrelevant the deaths of Nigerian environmental activists at the hands of the Nigerian dictatorship of

Table 11.2 Ethical product systems: a comparison of the pharmaceutical and energy sectors

Preconditions	*Pharmaceutical sector*	*Petroleum/fossil-fuels sector*
1. Agreement within key dominant companies within sector that common product and process standards are essential for consumer, government and stakeholder commitment.	Yes	Yes
2. A standardised cross-industry product-development procedure.	Yes	No
3. Fundamental commitment among all industry members to placing consumer interests as primary.	Yes	No
4. Government-regulated new-product approvals procedure.	Yes	No
5. Officially established product licensing procedure.	Yes	No
6. Product quality-assurance systems	Yes	Yes
7. Product contamination avoidance strategies.	Yes	Yes
8. Post-marketing product surveillance and reporting systems.	Yes	No
9. Regulated marketing procedures with systematic review of methods of dissemination.	Yes	No
10. Ongoing monitoring of product impacts and secondary effects, including side-effects for consumers.	Yes	No
Score	10 out of 10	3 out of 10

the day, were bitterly remembered by customers happy to switch to other brands. To its credit, Shell's response was to increase its environmental monitoring and verification procedures.

We see in these two cases, fossil fuel energy and ethical pharmaceuticals, two extreme types of environmentally linked industry. On the one hand, fossil energy has become generic. That is, its main products, gasoline and other car products, are now standardised and produced to a lowest attainable unit cost. So unprofitable are they that oil companies in their 'downstream' activities were by the late 1990s diversifying rapidly into high margin areas, particularly prepared foods and groceries. The price of raw crude had fallen to its lowest level in 30 years, as energy efficiency expanded in all economies and additional supplies came on stream from offshore Brazil and other new players.[6]

By contrast, drug companies have found methods of capturing intellectual property values within their products which can assure them during the period of patent, 20 years plus, a reliable stream of income. With high margins attached to the tiny percentage of successful products, demanding standards of environmental performance, as well as ethical marketing procedures, can embed themselves. High returns on capital occur, too.

The oil industry of coming decades can produce fuels which perform with extraordinarily higher efficiency than those which dominate modern economies. Indeed with Shell's announcement in 1998 that it aimed to establish itself as a major competitor in renewable and sustainable energy sources, the first essential step in market specialisation and product evolution fell into place.[7] Big oil had learned that there is more to product performance than its core utility value might suggest: customers increasingly have expectations of fair play in the origins of good products.

PRODUCT ETHICS AS A CORE CONSUMER VALUE: SPORTS SHOES

Sports footwear was being faced with some of these dilemmas in the late 1990s. Nike sportswear, a company founded in the 1970s by a group of running enthusiasts and professionals in Oregon, had by the 1990s achieved dominance over the young fashion markets around the world. Indeed, with the American government limiting shoe imports, Nike's main task for much of its life was not so much to sell shoes, but to ration supplies to the various American markets where demand continued to grow far ahead of capacity to supply.[8] Its shoes, initially developed for competing athletes and amateurs, increasingly became part of the fashion package for American and European youth. But its success and phenomenal profitability were based on a deep contradiction: Shoes designed in and sold in America were produced in low-cost factories throughout East and South East Asia, in conditions which critics equated to the sweatshops which for the most part had disappeared from the West.

Nike riposted that the sub-contractors producing their shoes paid above average wages for the regions in which they worked. While conditions in factories were perhaps demanding, most of the workers were much better off than in their previous lives as agricultural land labourers or unskilled labourers in urban shanty-towns. Shoe factories were thus genuine centres for regional development and personal change.

That said, there was no doubt that the profits were very large and the conditions far less than those expected in Western countries. However positive they were in local terms, there remained an *ipso facto* case of exploitation of poor producers by rich consumers. Reflecting international criticism, Nike at last in 1997 undertook to review and to establish benchmark minima for its offshore contractors. Soon thereafter, a recovery occurred in the eminence of the Nike brand among its previously loyal customers.

THE EVOLUTION OF CORPORATE CONTEXT CONCERNS: OUT OF THE ETHICAL GHETTO INTO THE STRATEGIC HEARTLAND OF CORPORATE STRATEGIC MANAGEMENT

Is business ethics a *non sequitur*? The ethical ghetto reflects those areas of strategic risk where corporate intelligence has not yet evolved consensus about how risk is to be priced and managed. What is clear is that history matters in issues of corporate strategy and governance. The key concerns of one decade become marginalised by changes in the business situation confronting the next. In the late 1990s, all businesses, but especially transnational corporations, are for the first time since the 1930s having to plan their businesses against a background which may bring about deflation and recession in many key markets.

Safety, health and environment, Paul Shrivastaya's famous SHE triumvirate, have imposed themselves on business life as core strategic concerns at different points in the modern industrial era.[9] As Table 11.3 suggests, critical moments in public life trigger legislation and customer demands which effectively require structural modification of prevailing strategic paradigms. From 'capitalism triumphant', first celebrated in the closing decades of the nineteenth century and then again in the Reagan–Thatcher era of the 1980s (e.g., see George Gilder's *Wealth and Poverty*), capitalism unleashed was expected to provide the highest net benefit for human development.[10] However, capitalism unleashed is unstable capitalism: it tends over time to self-destruct. By the end of the 1990s this reality was beginning to dawn.

Table 11.3 The evolution of SHE: corporate context and historical patterning

*Area of corporate strategic focus**	*Date parameters*	*Critical incident*	*From ethical ghetto*	*To strategic core*
Health (short and long-term personal hazard avoidance).	1850s onward.	Death and injury of children in factories and mines leads to child protection at work; hour limitations. Repeated in 1990s re sports clothing manufacturing in developing countries.	'To deprive children of the opportunity of work is to lower family living standards.'	Worker welfare as legal and moral responsibility of employers.
Safety (short-term personal hazard avoidance).	1960s	Product safety – cars, unsafe at any speed; Ralph Nader.	'We design products as well as we can for a price the consumer can afford.'	Consumer rights and product reliability/safety increasingly reflected in legislation, as well as competitive pressures.
Environment (long-term personal and societal hazard avoidance).	1970s.	1984 (Bhopal) – 1995 (North Sea and Nigeria).	'Not the concern of business...' 'Prosperity carries an inevitable price.'	Environmental care as an important secondary product; legislation and regulation of outputs; consumer expectations.

* Inclusive of employee, consumer, community; involving product, process and plants.

THE PROBLEM OF THE AHISTORICISM IN CONTEMPORARY BUSINESS STRATEGIC THEORY

Much of the powerhouse work of contemporary strategic business theory is developed without historical context.[11] For instance, the useful work of Porter over two decades like that of other applied economists seeks to relate to a series of postulated but non-articulated ideal types which can be called the typical corporate sector. The models are articulated as if history had never occurred and corporations themselves do not evolve in the context of society and values around them. Are corporations autonomous systems? Clearly they are not. Therefore, the pattern of development for corporations must be one in which their internal values and structures evolve in response to pressures from stakeholders around them.

THE NEW GLOBALISATION PARADIGM: FROM CAPITALISM TO COMMUNITARIANISM

In the later part of the decade, as the century waned, economists and policy-makers spoke much of 'The New Paradigm', a world without boundaries in which all markets were transparent and computer-driven tele-technology made instantaneous trades and values possible and accurate. In this vision of a global trading village, capitalism in its profit-seeking role seemed to be ultimately vindicated, for the knowledge of traders in currencies and commodities can instantly reward efficient and effective nations and companies and punish the reverse. If only it were so.

Another and contrasting framework became apparent following the collapse of the Thai baht in July 1997. The new global marketplaces revelled in currency speculation, one of the easiest places both to profit and to go broke. Currency funds, driven by large institutions, banks and pension funds as much as by speculators, fed endlessly on rumour and counter-rumour. They aimed to find numerous anomalies within the infinity of opportunities. In the end their profits mounted on the luckless backs of producers and economies dependent on volatile currency values for their day-to-day subsistence.

We can indeed project a New Paradigm for global capitalism. However, we reject the notion of an endless Global Casino but instead suggest a Global Community as the basis of the sustainable new paradigm. Within that ethical community, there are areas where instantaneous dealing will remain the economically most efficient method of enhancing global prosperity. However, in the last decade of the century, a new consensus about the limits in which these forms of trading can occur was needed. Otherwise, many commentators feared that countries like Brazil could follow in the footsteps of Thailand and Indonesia, suffering profound losses of stability and commodity value due to ungoverned speculation by international banking and institutional interests.

Table 11.4 The new globalisation paradigm: two contrasting approaches

Roles of:	*New paradigm A: Capitalism – the 'global casino'*	*New paradigm B: Communitarianism – the 'global community'*
Private enterprise	Unhindered maximisation of profit.	Negotiating sustainable profitability over at least two decades.
Government	Regulator and tax authority.	Facilitator, co-operating partner, regulator, taxation authority.
Multinational corporations	Amoral globalising maximisers.	Locally focused responsible community developers.
Non-govt. organisations	Counter-balance capitalism and ameliorate social impact of market economics.	Become commercially astute while remaining socially accessible.

A new global framework for regulation is outside the scope of this book. However, what is clear is that whatever the new arrangements are, they must take on board the ethical, environmental and human impacts of international business in ways which we are only beginning to explore. Table 11.4 outlines some of the issues arising for various sectors of the international economy in the two contrasting approaches.

A GLOBAL COMMUNITY PARADIGM FOR CORPORATE STRATEGY

There is little doubt that what is being sought requires huge changes in the way business conducts itself. Since the advent of industrial society at the end of the eighteenth century, industrial life has been driven by and in large part has depended on environmental pillaging. But it is a truism that the last two centuries of the second millennium after the birth of Christ have constituted an anomalous and abnormal interregnum.[12] The circumstances for wealth and poverty, even during these two centuries, have been distributed unequally.[13] Even today only a third of the six billion human beings alive can be said to live in comfort, with another third just about surviving and two billion always on the point of disaster. Prior to the industrial era and for tens of thousands of years, we humans extended our tenure over the face of the planet Earth, struggling and competing, but also acting as stewards of the many successful habitats offered for our species' undoubted supremacy.

With industrialism we developed for the first time technologies which can override the biospherical support system. That system had until then offered infinite support. How, given the powerful history of capitalist industrialism to the present day, are we to motivate hundreds of thousands of businesses, government agencies and other enterprises across the globe to change their way of thinking?

There are three interrelated strands which emerge from the analyses of this book and which summarise contemporary experience:

1. Look to the leaders of corporate life, the top companies and the Global 500 of *Fortune* magazine for leadership and innovation.
2. Re-design government agencies as change managers and partners in change.
3. Erect new international systems of discourse, dialogue and negotiation.

Behind these methods lie some unavoidable facts: change must come, if not in this generation, then in the second or third generations to come, and certainly sometime in the twenty-first century. If it does not, then the accumulation of atmospheric and biospheric toxins and consequent climatic instability and volatility will provide Gaia with its own self-correcting facility. The question is whether humankind will decide to mediate and strategise new corporate sustainabilities or to surrender passively to planetary destabilisation.

MAKING ENVIRONMENTAL SUSTAINABILITY THE NEW BENCHMARK FOR BUSINESS AND CORPORATE STRATEGY: SUMMARY OBSERVATIONS

Companies which are designed with environmental sustainability as their core value have been conceptualised and indicated for some time. However, it is fair to say that so far no major corporation has come into being which exemplifies in every aspect of production and product design full sustainability. The ethical pharmaceuticals may have one of the highest probabilities of attaining such a standard, as they already operate to rigorous and demanding cross-industry expectations on a number of fronts. Car companies are increasingly seeing their future as viable only under a sustainability banner (see Chapter 9).

Water companies, which are public corporations in Europe and publicly owned agencies in North America, have products which are fully recycled and designed to high levels of purity. They have every motivation to continue to develop in a good environmental setting.

The overview presented in this book shows many of the variegated versions of the sustainable corporation as it takes shape in the international milieu. From the macro-planetary process to the meso-micro terrestrial environment of businesses, there are now millions of managers and entrepreneurs working together with government and NGOs to create this new business ethic. The future of the globalised and environmentalised corporation is undoubtedly one of success, rather than marginalisation. It forms one of the great challenges of the new century and new millennium. This new business approach is no minor footnote to history. Rather, it is the harbinger of the future.

We project a proliferation of green businesses in coming decades, with improved social and environmental ethics becoming an accepted minimal standard. The intensity of communications in the global environment ensures rapid distillation of their best practices to every sector. Many businesses will becoming increasingly knowledge intensive and materially lighter, with telecommunications displacing goods transport in some cases. In short, we can have measured confidence in the future of business as a key ally to global sustainability, rather than the reverse.

Notes and References

Introduction, pp. 1–6

1. This discussion develops themes of earlier books: G. Ledgerwood, E. Street and R. Therivel, *The Environmental Audit and Business Strategy: A Total Quality Approach* (London: Pitman Financial Times, 1992); G. Ledgerwood, E. Street and R. Therivel, *Implementing an Environmental Audit* (Burr Ridge, III.: Irwin Professional Publishing, 1994); G. Ledgerwood (ed.), *Greening the Boardroom* (Sheffield, UK: Greenleaf, 1997). See also Arlene I. Broadhurst and Grant Ledgerwood, 'Environmental Diplomacy: Corporations, States and Non-Governmental Organisations in a World-wide Web of Interaction', *International Relations*, vol. XIV, no. 2 (August 1998) pp. 1–2; Grant Ledgerwood and Arlene Idol Broadhurst, 'Creating Technology-Based Enterprise Televillages: Post-Modern Regional Development Theory', *Cities, The International Journal of Urban Policy and Planning*, vol. 15, no. 1 (February 1999) 43–50.
2. D. Meadows, D. Meadows and J. Randers, *The Limits to Growth* (Cambridge, MA: MIT Press, 1972); D. Meadows *et al., Beyond the Limits to Growth* (New York: Chelsea Park, 1992).
3. Clive Cookson, 'World Must Look to Nature's Free Services', *Financial Times*, 17 February 1997, p. 5.
4. P. Sorsa, 'Competitiveness and Environmental Standards', Working Paper No. 1249 (Washington, D.C.: The World Bank, International Economics Department, 1994).
5. M. Scallon and M. Sten, 'Environmental Positioning for the Future: A Review of 36 Leading Companies in the Pacific Northwest', in G. Ledgerwood (ed.), *Greening the Boardroom* (1997), pp. 145–64.
6. Theory and context of corporate environmental governance are discussed in G. Ledgerwood (ed.), *Greening the Boardroom* (Sheffield: Greenleaf 1997), pp. 13–16. See also in the same volume, P. ten Brink and R Haines, 'World Environmental Industries: Market Drivers, Developments and Implications for Corporate strategies', pp. 43–55.

1 The environmentalisation of international business, pp. 9–25

1. Dr I. Rush, Director of Environmental Management, ICI plc. Address, Confederation of British Industry Conference on Corporate Environmental Strategy, London, May 1995.
2. P. Sorsa, 'Competitiveness and Environmental Standards', Working Paper No. 1249, The World Bank International Economics Department, Washington, D.C., 1994.
3. Doug Rushkoff, *Cyberia: Life in the Trenches of Hyperspace* (San Francisco: Harper, 1993, p. 21).

4. M. Hammer and J. Champy, *Re-engineering the Corporation: A Manifesto for Business Revolution* (New York: HarperCollins, 1993; London: Nicholas Brealey, 1994, pp. 50–56).
5. Napoleon Hill, *Think and Grow Rich* (Los Angeles: Wilshire Book Co., 1937 1st ed; UK: Thorsons, 1966 ed, pp. 100–14). There is a large and generally fugitive literature arising out of Hill's writings, widely read by would-be entrepreneurs across the English-speaking world. See, for instance, M. Maltz, *Psycho-Cybernetics: A New Technique for Using Your Subconscious Power* (Englewood Cliffs, N.J.: Prentice-Hall, 1960); D. J. Schwartz, *The Magic of Thinking Big* (North Hollywood, California: Wilshire Book Co., 1959); and A. Robbins, *Unlimited Power* (New York: Fawcett Columbine, 1986).
6. Editors, 'Energy Efficiency Survey', *Financial Times*, 28 November 1994, p. II.
7. Donella H. Meadows, Dennis L. Meadows, Jorgen Randers and William W. Behrens III, *The Limits to Growth: A Report for the Club of Rome's Project on the Predicament of Mankind* (Cambridge, MA: MIT Press, 1972); and D. H. Meadows *et al.*, *Beyond the Limits to Growth* (New York: Chelsea Park, 1992).
8. See Greg Easterbrook, *A Moment on the Earth: The Coming Age of Environmental Optimism* (UK: Penguin, 1996); and J. Vidal, 'Crisis, What Crisis?', *The Guardian*, 24 July 1996, pp. 4–5.

2 Global business innovation and the environment, pp. 26–41

1. This section is based on the important research of Monica Scallon and Mark Sten, reported in M. Scallon and M. Sten, 'Environmental Positioning for the Future' *Greener Management International Journal of Corporate Environmental Strategy & Practice*, vol. 13 (January 1996), pp. 49–65. Also in G. Ledgerwood (ed.), *Greening the Boardroom* (Sheffield, UK: Greenleaf Publishing, 1997) pp. 145–64.
2. S. Goldenberg, 'Silicon City Creaks Under the Strain of Its Success', *The Guardian*, (24) 10.
3. *Ibid.*
4. *Ibid.*
5. Manuela Saragosa, 'Freeport Mine Challenged on Environment', *Financial Times*, 20 April 1996, p. 3.
6. Grant Ledgerwood, worked with Dave Prince of Ohio University on a global arts network business plan during 1994–5.

3 Global trends and the corporate response, pp. 42–59

1. J. Stuart, 'World Review – April 1995', *Greener Management International, Journal of Corporate Environmental Strategy and Practice* (April 1995) pp. 6–33; Quotation from the *Financial Times* report, 4 April 1995.
2. D. Osborne and P. Gaebler, *Re-inventing Government* (New York: Addison-Wesley, 1992).

3. *The Economist*, 'A Survey of Development and the Environment', 21 March 1998, p. 4.
4. *The Economist*, *ibid*.
5. As reported on BBC, 'Our Own Correspondent', 24 May 1995.
6. Global Climate Coalition, 'The Global Climate Coalition Membership Backgrounder', 21 May 1998, p. 2.
7. *Ibid*.
8. Paul Brown, 'Change in Climate of Opinion', *The Guardian*, 'Society and Environment', 3 July 1996, pp. 4–5.
9. R. Baird, 'Stormy Weather Will Cost More', *The Guardian*, 'Society and Environment', 3 July 1996, pp. 4–5.

4 Environmental diplomacy and corporations in the worldwide web of interaction, pp. 60–82

1. A rather limited number of sources in French and English has served as the basis for the classical view of diplomacy: Abraham de Wicquefort, *L'Ambassadeur et ses Fonctions* (1679); Francois de Callieres, *De la maniere de negocier avec les Souverains* (1716); Charles de Martens, *Guide Diplomatique* (1821); A.H.L. Heere, *History of the Political System of Europe and Its Colonies* (1809); Albert Sorel, *L'Europe et la Revolution Française* (1885); Rt. Hon. Sir Ernest Satow, *A Guide to Diplomatic Practice* (1917, 1932, 1979); and Harold Nicolson, *Diplomacy* (London: Oxford University Press, 1939, 1950).
2. Rt. Hon. Sir Ernest Satow, ibid (1932 ed) p.1; Harold Nicolson, *ibid* (1950 ed) p. 15; Herbert Butterfield and Martin Wight (eds), *Diplomatic Investigations: Essays in the Theory of International Politics* (London: George Allen & Unwin, 1966), p. 182; Hedley Bull, *The Anarchical Society: A Study of Order in World Politics* (London: Macmillan, 1977), p. 13; G. R. Berridge and A. Jennings (eds), *Diplomacy at the UN* (London: The Macmillan Press Ltd., 1985), p. xvi; Alan James, 'Diplomacy', *Review of International Studies*, vol. 19, no. 1 (January 1993), p. 95.
3. See Nicolson, *Diplomacy* (1950 ed), pp. 14, 34, 59, 61, 129, 154, 162, respectively.
4. François de Callieres, quoted in Lauren., p. 27. For a more detailed discussion, see H.M.A. Keens-Soper and Karl W. Schweizer (eds), *The Art of Diplomacy* (Leicester University, 1983) and G.R.Berridge, *International Politics: States, Power and Conflict since 1945* (Brighton: Wheatsheaf 1987), p. 188.
5. Flassan, quoted in Satow, *A Guide to Diplomatic Practice*, p. 1.
6. J. Schmelzing, quoted in German in Satow, ibid, p. 1. Trans. Garth McCown.
7. de Cussy, quoted in French in Satow, *ibid*, p. 1. Trans. A. Broadhurst
8. de Martens, quoted in French in Satow, *ibid*, p. 1.
9. de Comte, quoted in French in Satow, *ibid*, p. 1.
10. Satow, *ibid*, p. 1.

11. H. Nicholson, *Diplomacy*, p. 15.
12. Ragner Numelin, *The Beginnings of Diplomacy: A Sociological Study of Inter-tribal and International Relations* (London: Oxford University Press, 1950), p. 14.
13. James Der Derian, *On Diplomacy* (Cambridge: Blackwell, 1987), p. 42.
14. G. R. Berridge, *Diplomacy: Theory and Practice* (Hertfordshire: Prentice Hall / Harvester Wheatsheaf, 1995), p. 1.
15. J. L. Holzgrefe, 'The Origins of Modern International Relations Theory', *Review of International Studies*, vol. 15, no. 1 (January 1989), 11–26.
16. See Alan James (1986) for a discussion of 'external' sovereignty; Robert Jackson (1986) for the idea of 'negative' sovereignty; and Arlene I. Broadhurst in David Haglund (ed.), *From Euphoria to Hysteria: Western European Security After the Cold War* (Boulder and London: Westview Press, 1993), pp. 58–63 for a discussion of variable sovereignty in assessing threats to security; James N. Rosenau, *The United Nations in a Turbulent World* (Boulder, Co: Lynne Rienner, 1992) and also his *Along the Domestic–Foreign Frontier* (Cambridge: Cambridge University Press, 1997), pp. 222 for a discussion of sovereignty as a continuum. For a much earlier challenge to territorial sovereignty, see John H. Herz, 'Rise and Demise of the Territorial State', *World Politics*, vol. 9, no. 4 (July 1957).
17. Robert H. Jackson and Carl G. Rosberg, 'Why Africa's Weak States Persist: The Empirical and Juridical Statehood', *World Politics*, vol. 35, no. 1 (October 1982).
18. Laurence E. Susskind, *Environmental Diplomacy: Negotiating More Effective Global Agreements* (New York and London: Oxford University Press, 1994, ch. 7). Susskind makes a similar point, along with many others, that co-operation is only possible when parties with competing interests have an opportunity to generate options for mutual gain; however he rejects the usefulness of the traditional convention-protocol approach and calls for a reform in the process of making treaties. See also Robert Axelrod, *The Evolution of Cooperation* (New York: Basic Books, 1984) and Robert O. Keohane and Elinor Ostrom, *Local Commons and Global Interdependence: Heterogeneity and Cooperation in Two Dimensions* (Boston: Center for International Affairs, Harvard University; Sage, 1995).
19. Sources for Table 4.2: Lynton Keith Caldwell, *International Environmental Policy: Emergence and Dimensions* (Durham, N.C.: Duke University Press, 1984); John E. Carroll (ed.), *International Environmental Diplomacy, The Management and Resolution of Transfrontier Environmental Problems* (Cambridge: Cambridge University Press, 1988); David Adamson, *Defending the World: The Politics and Diplomacy of the Environment* (London: I.B. Tauris & Co., 1990); Lynton K. Caldwell and Robert V. Bartlett (eds), *Environmental Policy, Transnational Issues and National Trends* (Connecticut: Quorum Books, 1997); U.S. Department of State, 'Environmental Diplomacy: The Environment and US Foreign Policy', *Annual Report on the Environment and Foreign Policy*, 1997.

20. R. Constanza (ed.), *Ecological Economics: The Science and Management of Sustainability* (New York: Columbia University Press, 1991); Jeroen C.J.M. van den Bergh, *Ecological Economics and Sustainable Development: Theory, Methods and Application* (Edward Elgar: UK, 1996) and Wallace E. Oates (ed.), *The Economics of Environmental Regulation* (Edward Elgar: Economists of the Twentieth Century Series, UK, 1996), pp. 675–740. See particularly the latter for a summary of the debate over effluent fees (taxation) vs. marketable permits to pay for pollution.
21. Peter Willetts (ed.), *The Influence of Non- Governmental Organisations in the U.N. System* (London: The David Davies Memorial Institute of International Studies and Hurst & Co., 1996), p. 56.
22. L.K. Caldwell (1987), *International Environmental Policy*, p. 28.
23. L. C. White, *International Non-Governmental Organizations* (New Brunswick: Rutgers University Press) 1951, pp. 30–31. Also discussed in Bill Seary, 'The Early History' in Peter Willetts, *The Influence*, p. 24.
24. Peter Willetts, 'Consultative Status for NGOs at the United Nations', in Peter Willetts, *The Influence*, p. 34.
25. Shanna L. Halpern, *The United Nations Conference on Environment and Development: Process and Documentation* (Academic Council on the United Nations System, Reports and Papers, 1993) no. 2, p. 19.
26. According to Craig Murphy, the BCSD was formed at the suggestion of Maurice Strong, Secretary-General of UNCED. See Albert Legault, Craig N. Murphy and W.B. Ofuatey-Kodje, *The State of the United Nations, 1992*, The Academic Council on the United Nations System, Reports and Papers, 1992, no. 3., p. 61.
27. World Business Council for Sustainable Development, *Annual Review 1996* (UK: Beacon Press, 1997), p. 1; *The Handbook of International Organizations*, 1996; and Stephan Schmidheiny, Rodney Chase and Livio DeSimone, *Signals of Change: Business Progress Towards Sustainable Development* (WBCSD Publications, Geneva, undated, 1998?), p. 4.
28. WBCSD, *ibid*, p. 17. The text refers to two regional councils, but the map indicates only one.
29. These examples are taken from WBCSD, *Signals of Change: Business Progress Towards Sustainable Development*, WBCSD, *Eco-efficient Leadership for Improved Economic and Environmental Performance*, 1966; WBCSD and The World Conservation Union (IUCN), *Business and Biodiversity: A Guide for the Private Sector*, June 1997.
30. The Kyoto Protocol to the UN Framework Convention on Climate Change, art. 13, no. 8, p. 15, *Green Globe Yearbook* (The Fritjof Nansen Institute, 1988). Article 13 provides admission for '…anybody or agency, whether national or international, governmental or non-governmental, which is qualified in matters covered by this Protocol.'
31. This was also a major theme developed by Dr Andrew Dlugolecki of General Accident in a public lecture entitled '*Environmental Risk Management: What influences business to make environmentally responsible decisions*', delivered at the London School of Economics, 17 March 1998.
32. WBCSD newsletter, *Sustain*, Issue 3, April 1997, pp. 1–3.

33. *Shell and the Environment* (London: Group Public Affairs, Shell International Petroleum Co., 1995), p. 13.
34. For an excellent summary of the growth of environmental NGOs, see Sally Morphet, 'NGOs and the Environment', in Peter Willetts, *The Influence*, ch. 5.
35. See Grant Ledgerwood, 'Environmental Stewardship of the Planet', in Grant Ledgerwood (ed.), *Greening the Boardroom: Corporate Governance and Business Sustainability* (Sheffield: Greenleaf Publishing, 1997), p. 22. Also Jan-Olaf Willums and Ulrich Goluke, *From Ideas to Action: Business and Sustainable Development: The ICC Report on the Greening of Enterprise* (Oslo: ICC Publishing and Ad Ad Notam Gyldendal) 1992.
36. Titus Moser and Damian Miller, 'MNCs' Impacts on the Environment and Communities in the Developing World: A Synthesis of the Contemporary Debate', *Greener Management International*, 19 (Autumn 1997), pp. 40–51.
37. Trevor Russell 'World Review', *Greener Management International*, 19 (Autumn 1997), pp. 9–10.
38. *Shell and the Environment* (London: Group Public Affairs, Shell International Petroleum Co.), 1995, p. 13; and Shell UK *Environmental Report 1995* (London: Shell U.K. Limited), 1995, p. 19.
39. *Health, Safety and Environment Report*, 1997 (The Hague: Royal Dutch/Shell Group of Companies, 1997), p. 30; Murray Jones, 'The Role of Stakeholder Participation: Linkages to Stakeholder Impact Assessment and Social Capital in Camisea, Peru', *Greener Management International*, 19, pp. 88–97. Sources for Table 4.3: 1995–1997 editions of *Greener Management International*; official publications from Shell UK and the WBCSD previously cited.
40. Jennifer B. Shulman, 'The North American Free Trade Agreement (NAFTA): What it is and how it came to be', in Madeleine O. Hosli and Arild Saethe (eds), *Free Trade Agreements and Customs Unions: Experiences, Challenges and Constraints* (Brussels: European Commission and Mastricht, the Netherlands: The European Institute of Public Administration), 1997. p. 209; pp. 216–18.
41. Nick Robbins and Sarah Roberts, 'Trade Opportunities for Developing Country Producers', *Greener Management International*, 19 (Autumn 1997), pp. 53–66.

5 Greenpeace: case study of an NGO multinational enterprise, pp. 83–108

1. As members of Grant Ledgerwood's MBA Strategic Management course at the University of Greenwich, Joanne Clarke, Jillian Cross, John Griffiths, Mervyn Holt, Gordon Hardwick and John Kent contributed the initial case study of this chapter. We are grateful for Gordon Hardwick's help in gaining permission to use parts of the case report. Given the substantial revision in the case study, they are not responsible for its content.
2. Kevin Stairs and Peter Taylor, 'NGOs and Legal Protection of the Oceans', in Andrew Hurrell and Benedict Kingsbury (eds), *The*

International Politics of the Environment (Oxford: Clarendon Press), 1992, p. 113.

3. *The Economist* (19 August 1995), pp. 65–6.
4. Oliver August, 'Greenpeace – Protesters Reluctant to Tap into Reserves', *The Times*, London, 7 December 1995.
5. Matt Ridley, 'Greenpeace – A Cause Driven by Cashflow', *Sunday Telegraph*, London, 10 Sept. 1995.
6. Charles Hymans and Steven Haynes, 'Greenpeace's Millions Hidden Around the World', *Sunday Times*, London, 22 Oct 1995.
7. 'Greenpeace Stories', Interview by Michael Friedrich of Jim Bohlen and Dorothy Metcalf, Greenpeace web site: www.greenpeace.org See also 'Greenpeace – Twenty-five Years of Environmental Activism', 'Booklist', Greenpeace web site, and Kate Kelland, 'Greenpeace Co-founder Revisits Early Campaign Days', 29 September 1995, Reuters News Service, Amsterdam.
8. Greenpeace, '25 Years of Environmental Activism', Greenpeace web-site document.
9. Quoted in *The Turning of the Spar*, published by Greenpeace, London, 1998 (ISBN871532 99X), p. 215.
10. 1997 Annual Report, Greenpeace world wide pro forma summary financial statements, downloaded from the Greenpeace web site on 8 November 1998, p. 1.
11. Greenpeace publication 'Overview', March 1997, L3, P28.
12. *The Economist*, 'Greenpeace Means Business', 19 August 1995, pp. 65–6.
13. The eight are Argentina, Brazil, Central America, Chile, China, Latin America, Mexico, Tunisia. Central and Eastern European offices are not considered to be in less developed countries.
14. Peter Melchett, 'Greenpeace Not Principally a Charity', Letters, *Financial Times*, 21 September 1995.
15. Charles Hymas and Steven Haynes, 'Greenpeace's Millions Hidden Around the World', *Sunday Times*, London, 22 October 1995.
16. James Meikle, 'Greenpeace Dumps Two More of Its Warriors', *Guardian*, London, 27 January 1996.
17. In research conducted in 1994, there were four Greenpeace companies registered in the UK and one Trust. The UK accounts were exempt from full disclosure through provisions of the FRS1 cash flow and from the need for consolidated accounts by S.248 of the Companies Act, 1985. Many of the global accounts were registered in Holland and are exempt from full disclosure due to their tax-exempt status. Therefore funds can be channelled between accounts secretively. In 1998, the Greenpeace National Office accounts remain unaudited as is the world-wide pro forma summary financial statements.
18. *Financial Times*, London, 21 June 1995.
19. 'Auditor's Report', Financial Statements, Greenpeace International pro forma summary statements for the years ended 31 December 1997 and 1996, at www.greenpeace.org/report97/finance2, p. 1.
20. Greenpeace International, *1997 Annual Report*, 'Greenpeace International pro forma summary financial statement, years ended 31 Decem-

ber 1997 and 1996, including KMPG Accountants, Amsterdam, Auditor's Report, 15 July 1998. Greenpeace: www.greenpeace.org/report97/finance2., downloaded 9 November 1998.

21. For example, Hymas and Haynes reported that £400,000 was paid to four departing Greenpeace bosses, some of whom had been asked to leave quietly. See Charles Hymas and Steven Haynes, 'Greenpeace's Millions Hidden Around the World', *Sunday Times*, London, 22 October 1995.
22. Interviews were conducted with Greenpeace London representatives about their use of telematics.
23. Robert Rice, 'Greenpeace Steps Up Campaigns', *Financial Times*, London, 11 September 1995.
24. Philip Kotler, Donald H. Haider, Irving Rein, *Marketing Places: Attracting Investment, Industry and Tourism to Cities, States and Nations* (New York: The Free Press, Macmillan, Inc., 1993).
25. Gerry Johnson and Kevan Scholes, *Exploring Corporate Strategy* (Hempel Hempstead: Prentice Hall Europe, 1999).
26. Bronwen Maddox, 'Rubber Suits Turn the Tide for Greenpeace', *Financial Times*, London, 21 June 1995.
27. Matt Ridley, *Daily Telegraph*, 10 September 1995.

6 The metropolitan corporate centre: reinvention of the medieval city-state, pp. 111–41

1. A European Community Directive, implemented in the UK by Circular 15/88, instituted major project environmental assessment across Europe. EIA in the European context builds on twenty years of American experience. Environmental impact assessment has been a requirement for major federal land development in the US since the early 1970s. Its application to private sector development is determined by state law which is more rigorous in some states (New York, California, Colorado, Oregon) than in others, particularly rural ones.
2. See Environmental Resources Ltd. and Montagu Evans *et al.*, *London Regeneration Consortium Proposals, King's Cross Railway Lands*, London, 1989. We have relied on the published data and materials relating to the King's Cross Proposals of the London Regeneration Consortium, including the Environmental Statement relating to that proposal.
3. The environmental assessment for the King's Cross Team was conducted as a scoping exercise by the co-author, Grant Ledgerwood, and Elizabeth Street. For an earlier description of this project, see Grant Ledgerwood, E. Street, R. Therivel, *Implementing and Environmental Audit – How to Gain Competitive Advantage through Quality and Environmental Management* (Burr Ridge, Illinois: Irwin Professional Publishing), 1994, and Grant Ledgerwood, Ian Haywood, Norman Sheppard and Elizabeth Street, 'Environment Assessment as a Real Estate Management Protocol', *Cities*, vol. 11, no. 3 (1994), pp. 181–94. The EuroCity case study was delivered as a conference paper, entitled 'Environmental Issues and Imperatives for the 21st Century', at City University of New York in April 1995.

4. Michael Parkes, *King's Cross Railwaylands: People's Brief* (London: King's Cross Railwaylands Group, KXT/KXRLG, December 1990). Michael Parkes's advice on local community development was valuable in this case-study research. All the investigators were specialists and observers on London planning prior to the King's Cross programme and all were involved in the King's Cross team (KXT) set up by Martin Clarke and Co. to develop a lower-scale alternative to the LRC competition winner. On the KXT, Sheppard and Haywood were initiators and team leaders; Street and Ledgerwood acted latterly as environmental assessors and consultants. The rediscovery of community-based architecture for great cities was piloted by the American urban specialist, Jane Jacobs. See her *Life and Death of Great American Cities*, 1963.
5. A major series of research studies on King's Cross Railway Lands and adjacent communities was developed at University College, London. See Michael Edwards, Ellen Leopold, Michael Geddes, *King's Cross Railway Lands Redevelopment: Second Report* (London: University College, Bartlett School of Architecture and Planning, 1990).
6. D. Lichfield and N. Lichfield, *Community Impact Evaluation: Principles and Practice* (London: University College Press, 1988).
7. Protocol is defined as a procedure for achieving mutually acceptable answers in situations where contenton and contestation – adversarial settings – prevail.
8. Some of these issues are examined in Ledgerwood *et al.*, *The Environmental Audit and Business Strategy* (London: Financial Times Pitman Series, 1992), ch. 5. See also G. Ledgerwood, and E. Street, 'Corporate Strategy and Environmental Sustainability: Establishing an Environmental Market Ethos to Gain Competitive Advantage', *GMI Journal of Corporate Environmental Strategy and Practice*, vol. 1, no. 3 (July 1993), pp. 41–50.
9. See G. Ledgerwood, *Urban Innovation: London's Docklands* (Aldershot UK: Gower Publishing, 1985) ch. 2, for further consideration of planning paradigms as decision contexts.
10. H. Simon, *The Art of Decision* (New York: Macmillan, 1957).
11. Martin Jacques, 'Where the Future is Outrageous', *The Guardian*, 22 March 1997, p. B5.
12. The concept of the 'enterprise televillage', a registered trademark, has been developed by a specialist partnership based in the University of Greenwich Business School. With the assistance of a Housing Corporation UK Innovation Grant (1998), several specialist advisory committees in the Kent region have been working on applying the concept to urban regeration locations. A parallel research project is based with David Prince and Glenn Corlett in Ohio University's College of Business, Athens, Ohio, USA. For additional discussion see G. Ledgerwood and A. Broadhurst, 'Creating Technology-Based Enterprise Televillages: Towards a Post-modern Regional Development Theory', *Cities*, vol. 16, no. 1 (February 1999) 43–50.

7 The environmental entrepreneur: personality and culture, pp. 142–68

1. M. Webber, 'The Non-Place Urban Realm and the New Metropolitan Structure', in Donald Foley (ed.), *Explorations in Urban Structure* (Philadelphia: University of Pennsylvania Press, 1967), pp. 79–153.
2. Napoleon Hill, *Think and Grow Rich* (Los Angeles: Wilshire Book Co., 1935.)
3. Joshua Cooper Ramo, 'The Three Marketeers', Interview with Greenspan, *Time*, 15 February 1999, p. 25.
4. Maxwell Maltz, *PsychoCybernetics: A New Technique for Using Your Subconscious Power* (North Hollywood, CA: Wiltshire Publishing, 1960 and subsequent editions). Maltz, a physician, discovered empirically the power of mental images in overcoming mental and physical disability and led in the development of an influential body of work, as well as extensive emulators and researchers, in the decades since this book first appeared.
5. Karlin Lillington, 'Walk on the Wired Side', *The Guardian Online*, Thursday, 14 January 1999, p. 2.
6. *The Guardian*, News Section (Friday, 15 January 1999), p. 3.
7. The psychologist, Dr. Harriet Mann, developed in her unpublished manuscript of 1972, *The Pooh Stakehold* (Cambridge MA), a provocative, if non-operational concept for comparing modal personalities. Her work anticipated the findings of more recent times that there are many acceptable and admirable versions of 'normality' and excellence in human organisations.
8. Peter Briggs Myers and Katherine D. Myers, *Myers-Briggs Type Indicator, 1997* (Consulting Psychologists Press, Inc., Palo Alto, CA, 1997).
9. S. Larsen and J. Rogers, *Silicon Valley Fever* (New York: Doubleday, 1984). For a more recent overview, see Survey-Technology Industry, *Economist*, 29 March 1997.
10. R. Meredith Belbin, *Management Teams: Why They Succeed or Fail* (London: Heinemann, 1981).
11. *The Guardian*, London, 19 January 1999, p. 12.
12. Martin J. Wiener, *English Culture and the Decline of the Industrial Spirit (1850–1980)*. (Cambridge University Press, 1981); T. Forester, *High-Tech Society* (Blackwell, 1987).
13. Belbin, *Management Teams*, pp. 65ff.

8 Corporate innovation in the fast lane: environmental technologies and planning the new industrial metabolism, pp. 169–88

1. Energy Efficency Survey, *Financial Times*, 28 November 1994.
2. Automobile Association data cited in *The Guardian*, 10 February 1999, p. 15.
3. Duvall Smith (ed.), *Business and the Environment: Implications of the New Environmentalism* (London: Paul Chapman Publishing, 1993).

4. 'EV Facts & Myths', Electric Vehicle Information Network, 1997 website of the Electric Vehicle Association of Greater Washington, DC (http://www.radix.net/, 1/27/99), p. 3.
5. Gary Starr, 'The Shocking Truth About Electric Cars', Solarelectric, (http://metalab.unc.edu/ pp. 1–3.
6. EV News and Events, Electric Vehicle Information Network, p. 1. The suggested retail price for the Altra is US$50,999, but the costs are expected to decline over the next few years. See also www.nissan-usa.com and www.nissan-na.com for details.
7. AFX News Ltd (AX) 21/12/98; reprinted on www.radix.net/, p.3.
8. C. Aubrey, 'The Light Fantastic: Solar Housing', *Guardian*, Society, 16 August 1995, p. 4.
9. Aspen Institute, *Factor 4* (Colorado: Aspen Institute, 1995).
10. *Newsweek*, 22 February 1999, p. 6.
11. For example, see C. Webb, 'Dial the Kettle', *Sunday Times*, 5 January 1999, p. 4. Thanks to Mary Spence for drawing our attention to this article.
12. Personal communication with Grant Ledgerwood.
13. Most of these examples come from David Pearson, *Earth to Spirit: The Search for Natural Architecture* (London: Gaia Books Ltd, 1994) (USA edition, 1995).
14. Paolo Soleri, *Arcology: The City in the Image of Man* (Cambridge, Mass.: MIT Press, 1969).
15. Richard Register, *Ecocity Berkeley: Building Cities for a Healthy Future* (Berkeley, California: North Atlantic Books, 1987).
16. David Pearson, *Earth to Spirit: In Search of Natural Architecture* (San Francisco: Chronicle Books, 1995), p. 147; also David Pearson, *The Natural House Book* (New York: Simon & Schuster, 1989). Gaia is a concept of the earth which views it as a living organism, made up of components which can be compared to human organs. It was expanded by James Lovelock to the idea that the earth is a self-sustaining system where the evolution of organisms is symbiotically related to the evolution of the environment.
17. See M. Prest and S. Caulkin, 'Where There's Muck: Tarmac', *The Observer*, Business, 26 May 1995, p. 4.
18. Prest and Caulkin, *ibid.*
19. Paul Brown, *The Guardian*, 8 February 1999, p. 5. The most recent data available in a coherent form from the Environmental Agency was for 1996. Clearly the EP needs to improve its data collection methods if the 'naming and shaming' strategy of the environmental groups is to succeed.
20. This analysis owes much to Jorm Heinonen, 'Small Firms and Environmental Technology: An International Perspective', *GMI Journal*, vol. 8 (1994), pp. 37–47. See also Patrick ten Brink and R. Haines, 'World Environmental Industries: Market Drivers, Developments and Implications for Corporate Strategies', *GMI Journal*, vol. 8 (1994), pp. 26–36.

9 Nations and cities: sustainability models, pp. 169–206

1. *Economist*, 29 July 1995.
2. *The Guardian*, 15 January 1999, p. 3.
3. Clark Wieman, 'Downsizing Infrastructure', *Technology Review*, May 1996, pp. 48–55.
4. See Clark Wieman, *ibid.*, for an expanded discussion of these improvements.
5. Clark Wieman, *ibid.*
6. See Grant Ledgerwood, *Urban Innovation: London's Docklands, 1968–1984* (Aldershot: Gower, 1985) for an analysis of this process.
7. Susan Brownhill, *London's Docklands: Another Great Planning Disaster?* (London: Pluto Press, 1988).
8. *Economist*, 19 August 1995.
9. See Chapter 10 for a discussion of the problems in the garment sector which have occurred as a result of supply-chains relocating.
10. This profile, up to 1995, relies on David Tannenbaum's field research. See D. Tenenbaum, 'The Greening of Costa Rica', *Technology Review*, October 1995, pp. 42–53.
11. Tenenbaum, *ibid.*, p. 52.
12. Virginia Morell, 'In Search of Solutions', *National Geographic*, vol. 195, no. 2. (February 1999) p. 84.
13. *The Guardian*, Home News, Eurostat data, 10 February 1999, p. 7.

10 Business ethics: national and international dimensions of social responsibility, pp. 209–34

1. Sir Geoffrey Chandler, 'The New Corporate Challenge', *Time*, 1 February 1999, p. 68.
2. The sources used for this summary are: John Adair, *Management and Morality: The Problems and Opportunities of Social Capitalism* (Farnborough, Hants: Gower, 1980); George C. S. Benson, *Business Ethics in America* (Aldershot: Gower distributor for Lexington Books, 1982); John R. Boatright, *Ethics and the Conduct of Business* (London: Prentice-Hall International, 1997); Peter W. F. Davies (ed.), *Current Issues in Business Ethics* (London: Routledge, 1997); John Donaldson, *Business Ethics: A European Casebook* (London: Academic, 1992); Thomas Donaldson, *The Ethics of International Business* (New York: Oxford University Press, 1989); Thomas Donaldson and Thomas W. Dunfee, *Ethics in Business and Economics* (Aldershot: Ashgate, 1997); George Toyder, *The Just Enterprise: A Blueprint for the Responsible Company* (London: Adamantine Press, 1993); Jennifer Jackson, *An Introduction to Business Ethics* (Oxford: Blackwell, 1996); Alan Kitson and Robert Campbell, *The Ethical Organisation: Ethical Theory and Corporate Behaviour* (Basingstoke: Macmillan Business, 1996); Paul Minus (ed.), *The Ethics of Business in a Global Economy* (Boston and London: Kluwer Academic Publishers, 1993); Ricardo Petrella, *Limits to Competition: Towards a New Social Contract* (Cambridge, Mass: MIT Press, 1995);

Elaine Sternberg, *Just Business: Business Ethics in Action* (London: Little, Brown & Co., 1994); Robert C. Solomon, *Ethics and Excellence: Co-operation and Integrity in Business* (New York: Oxford University Press, 1993); R. C. Solomon, *The New World of Business: Ethics and Free Enterprise in the Global 1990s* (UK: Littlefield Adams Quality Paperbacks, 1994); Manuel Velasquez, *Business Ethics: Concepts and Cases* (London: Prentice-Hall International, 4th edn, 1998); Thomas W. Whipple and Dominic F. Swords, *Business Ethics Judgements: A Cross-Cultural Comparison Between the USA and the UK*, Working Paper Series (Henley-on-Thames: Henley, The Management College, 1991).

3. See Thomas W. Whipple and Dominic F. Swords, *Business Ethics Judgements: A Cross-Cultural Comparison Between the USA and the UK*. Working Paper Series (Henley-on-Thames: Henley, The Management College, 1991). This fascinating study of 319 students enrolled in business courses in the US and Britain concluded that 'respondents in the USA are more ethical than their UK counterparts' (p. 8) and that 'females are more ethical than males' (p. 8). They also concluded that 'cross-cultural differences are less important than gender differences in explaining the students' ethical judgements' (p. 9).
4. Kevin T. Jackson, 'Globalizing Corporate Ethics Programs: Perils and Prospects', in Marilynn Fleckenstein, Mary Maury, Laura Pincus and Pat Primeaux (eds), *From the Universities to the Marketplace: The Business Ethics Journey*, The Second Annual Vincentan Conference (Kluwer Academic Publishers, 1997), pp. 17–25. Jackson identified the first eight of these measures; the rest have been added.
5. Jean Jacques Servan-Schreiber, *Le défi américain* (London: Hamish Hamilton, 1968); Richard Barnet and Ronald Mueller, *Global Reach: The Power of Multinational Corporations* (New York: Simon & Schuster, 1974), and Raymond Vernon, *Sovereignty at Bay* (New York: Basic Books, 1971) are good examples of the early works. More recently, Frederick Clairmonte and John Cavanagh, 'Transnational Corporations and Global Markets: Changing Power Relations', *Trade and Development: An UNCTAD Review*, 4 (Winter), pp. 149–82; Joan Edelman Spero, *The Politics of International Economic Relations* (New York: St Martin's Press, 1990, 4th edn); Theodore H. Moran (ed.), *Multinational Corporations* (Lexington, Mass.: Lexington Books, 1985).
6. John Donaldson (ed.), *Business Ethics: A European Casebook* (London: Academic Press, 1992), pp. 63–5.
7. *The Guardian*, Home News Section, 23 February 1999, p. 10; also www.simonjones.org.uk
8. *The Guardian*, 'International News', London, 15 January 1999, p. 15.
9. *The Guardian*, 'Finance and Economics', London, 23 February 1999, p. 20.
10. The hiatus between action in 1948 and 1976 was due in part to the Cold War period in which the Soviet bloc and the Western States bloc could not reach agreement on the inclusion of economic, social and cultural rights on the one hand and civil and political rights on the other. The dispute was eventually resolved by devising two separate covenants to which individual states could adhere. Since the collapse of the Soviet Union and the cataclysmic changes in the political topography of the

European landscape in 1989, the issues of the dispute have disappeared and the Universal Declaration of Human Rights, including both covenants, is now widely accepted.

11. Shell UK Ltd., *Report to Society* (May 1998), p. 2.
12. Shell, *ibid*, p. 9.
13. Shell, *ibid*, p. 8.
14. Barry B. Hughes, *Continuity and Change in World Politics: The Clash of Perspectives*, 2nd ed (Englewood Cliffs, New Jersey: Prentice-Hall, 1994), p. 360.
15. Shell, *Report*, p. 27.
16. Figures from the US Fair Trade Federation cited in Mick Blowfield, 'Ethical Trade and Development – The Scope, Functioning and Benefit of Ethical Trade', unpublished paper, Natural Resources and Ethical Trade Programme, Natural Resources Institute, University of Greenwich, Chatham Maritime, Kent. See also the NRI website: http://www.nir.org/NRET/nret.htm
17. See Mick Blowfield, *ibid*, for an incisive analysis of ethical trade as an approach to economic and social development.

11 Globalisation and the environmentalised corporation: present exemplars and future prospects, 235–47

1. We have attempted here to characterise the complexity of the pharmaceutical industry's product development and marketing cycle in relation to the themes of this book. See R. F. White and S. Fraley, 'Imperfect Competition, Price Fairness and the Pharmaceutical Industry', *OnLine Journal of Ethics*, 1997 (/DePaul University, Institute of Business and Professional Ethics, September 1998): http://www.depaul.edu/ethics/
2. See Larry Churchill, *Rationing Health Care in America: Perceptions and Principles of Justice* (Notre Dame: University of Notre Dame Press, 1987); James Harvey Young, *The Toadstool Millionaires: A Social History of Patent Medicines in America Before Federal Regulations* (New Jersey: Princeton University Press, 1972); Roberta J. Apfel and Susan Fisher, *To Do No Harm; DES and the Dilemmas of Modern Medicine* (New Haven: Yale University Press, 1984).
3. See Mary Graham, 'The Quiet Drug Revolution', *The Atlantic*, issue no. 267 (Jan. 1991).
4. In this respect the Fair Trade agribusiness organisation has in English-speaking countries during the 1990s driven hard to establish better producer incomes for internationally traded commodity products like coffee and tea. An NGO-driven Western concept, it has provided what some see as alien major interventions in indigenous agricultural marketing systems. They have on the other hand had no systems for distribution and payment.
5. See G. Ledgerwood, 'New Corporate Governance Paradigms for Transnational Enterprises – Shell and Strategic Futures for Big Oil', *Corporate Governance – An International Review*, vol. 6, no. 4 (October 1998), pp. 269–7.

6. Oil industry professionals assured GL that even with Brent Oil at $13 per barrel there are still considerable profit margins, as oil extraction costs, even in North Sea oil fields, continue to fall year after year.
7. Interested readers may wish to refer to the exceptionally useful web-page of DePaul University's Institute of Business and Professional Ethics: www.depaul.edu/ethics/
8. Personal communication to GL, July 1987, Ron Buel, USA marketing manager, Nike headquarters, Beaverton, Oregon.
9. Paul Shrivastaya, 'Greening of Business', in Denis Smith (ed.), *Business and the Environment: Implications of the New Environmentalism* (London: Paul Chapman Publishing), pp. 27–39.
10. George Gilder, *Wealth and Poverty* (New York: Basic Books, 1981).
11. See Gary Hamel and C. K. Prahalad, *Competing for the Future* (Boston: Harvard Business School Press, 1996). Also Michael E. Porter, *The Comparative Advantage of Nations* (London: Collier Macmillan, 1989).
12. David Easterbrook, *A Moment in the Earth: The Coming Age of Environmental Optimism* (New York: Viking, 1995).
13. See David Landes's eloquent and elegiac comment on the conditions through which nations attain industrial wealth and poverty in the modern age. D. Landes, *The Wealth and Poverty of Nations* (London: Heinemann, 1997).

Bibliography

ADAIR, JOHN, *Management and Morality: The Problems and Opportunities of Social Capitalism* (Farnborough, Hants: Gower, 1980).

ADAMS, R., J. CARRUTHERS and S. HAMILL, *Changing Corporate Values: A Guide to Social and Environmental Policy and Practice in Britain's Top Companies* (London: Kogan Page, New Consumer, 1991).

ADAMSON, DAVID, *Defending the World: The Politics and Diplomacy of the Environment* (London: IB Tauris & Co., 1990).

AFX News Ltd (AX) 21 Dec. 98; reprinted on www.radix.net/

APFEL, ROBERTA J. and SUSAN FISHER, *To Do No Harm: DES and the Dilemmas of Modern Medicine* (New Haven: Yale University Press, 1984).

Aspen Institute, *Factor 4* (Colorado: Aspen Institute, 1995).

AUBREY, C., 'The Light Fantastic: Solar Housing', *Guardian*, Society, 16 August 1995, p. 4.

AUGUST, OLIVER, 'Greenpeace – Protesters Reluctant to Tap into Reserves', *The Times*, London, 7 December 1995.

AXELROD, ROBERT, *The Evolution of Cooperation* (New York: Basic Books, 1984).

BAIRD, R., 'Stormy Weather Will Cost More', *The Guardian*, 'Society and Environment', 3 July 1996, pp. 4–5.

BARNET, RICHARD and RONALD MUELLER, *Global Reach: The Power of Multinational Corporations* (New York: Simon & Schuster, 1974).

BBC, *Our Own Correspondent*, 24 May 1995.

BEAUCHAMP, TOM L. and NORMAN E. BOWIE (eds), *Ethical Theory and Business* 4th ed, (Englewood Cliffs: Prentice-Hall, 1993).

BELBIN, R. MEREDITH, *Management Teams: Why They Succeed or Fail* (London: Heinemann, 1981).

BENSON, GEORGE C. S., *Business Ethics in America* (Aldershot: Gower distributor for Lexington Books, 1982).

BERRIDGE, G. R., *International Politics: States, Power and Conflict since 1945* (Brighton: Wheatsheaf, 1987).

BERRIDGE, G. R., *Diplomacy: Theory and Practice* (Hertfordshire: Prentice-Hall – Harvester Wheatsheaf, 1995).

BERRIDGE, G. R. and A. JENNINGS (eds), *Diplomacy at the UN* (London: Macmillan Press Ltd., 1985).

BLOWFIELD, MICK, 'Ethical Trade and Development – The Scope, Functioning and Benefit of Ethical Trade', unpublished paper, Natural Resources and Ethical Trade Programme, Natural Resources Institute, University of Greenwich, Chatham Maritime, Kent.

BOATRIGHT, JOHN R., *Ethics and the Conduct of Business* (London: Prentice-Hall International, 1997).

BRIGGS MYERS, PETER and KATHERINE D. MYERS, *Myers – Briggs Type Indicator, 1997* (Palo Alto, Calif.: Consulting Psychologists Press, Inc., 1997).

BROADHURST, ARLENE IDOL (contributing editor), *The Future of European Alliance Systems: NATO and the Warsaw Pact* (Boulder, Colo.: Westview, 1982).

BROADHURST, ARLENE IDOL,'Forward to the Past: A Long View of The Long Peace' in David Haglund (ed.), *From Euphoria to Hysteria: Western European Security After the Cold War* (Boulder and London: Westview Press, 1993) pp. 58–63.

BROADHURST, ARLENE IDOL (with GRANT LEGERWOOD), 'Creating Technology-Based Enterprise Televillages: Post-Modern Regional Development Theory', *Cities, The International Journal of Urban Policy and Planning*, vol. 16, no.1 (February 1999) pp. 43–50.

BROADHURST, ARLENE I. and GRANT LEDGERWOOD, 'Environmental Diplomacy: Corporations, States and Non-Governmental Organisations in a Worldwide Web of Interaction', *International Relations*, vol. XIV, no. 2 (August 1998) pp. 1–2.

BROWN, PAUL, 'Change in Climate of Opinion', in *The Guardian*, 'Society and Environment', 3 July 1996, pp. 4–5.

BROWNHILL, SUSAN, *London's Docklands: Another Great Planning Disaster?* (London: Pluto Press, 1988).

BULL, HEDLEY, *The Anarchical Society: A Study of Order in World Politics* (London: Macmillan, 1977).

Business in the Environment (UK), *City Analysts and the Environment: A Survey of Investment Houses and Analysts in the City of London* (London: Extel, 1994).

BUTTERFIELD, HERBERT and MARTIN WIGHT (eds), *Diplomatic Investigations: Essays in the Theory of International Politics* (London: George Allen & Unwin, 1966).

CALDWELL, LYNTON KEITH, *International Environmental Policy: Emergence and Dimensions* (Durham, N.C.: Duke University Press, 1984).

CALDWELL, LYNTON K. and ROBERT V. BARTLETT (eds), *Environmental Policy, Transnational Issues and National Trends* (Connecticut: Quorum Books, 1997).

Camden London Borough Council, *King's Cross Railway Lands: Brief for Environmental Assessment* (September 1988).

Camden London Borough Council, *Brief for Social Impact Study, King's Cross Railway Lands* (February 1990).

CARROLL, JOHN E. (ed.), *International Environmental Diplomacy: The Management and Resolution of Transfrontier Environmental Problems* (Cambridge: Cambridge University Press, 1988).

CHANDLER, SIR GEOFFREY, 'The New Corporate Challenge', *Time*, 1 February 1999, p. 68.

CHURCHILL, LARRY, *Rationing Health Care in America: Perceptions and Principles of Justice* (Notre Dame: University of Notre Dame Press, 1987).

CLAIRMONTE, FREDERICK and JOHN CAVANAGH, 'Transnational Corporations and Global Markets: Changing Power Relations', *Trade and Development: An UNCTAD Review*, vol. 4 (Winter), pp. 149–82.

CONSTANZA, R. (ed.), *Ecological Economics: The Science and Management of Sustainability* (New York: Columbia University Press, 1991).

COOKSON, CLIVE, 'World Must Look to Nature's Free Services', *Financial Times*, 17 February 1997, p. 5.
CRAIG, GORDON, 'On the Nature of Diplomatic History: The Relevance of Some Old Books', in PAUL GORDEN LAUREN (ed.), *Diplomacy: New Approaches in History, Theory and Policy* (New York: Free Press, 1979) p. 26.
DAVIES, PETER W. F. (ed.), *Current Issues in Business Ethics* (London: Routledge, 1997).
DE CALLIERES, FRANCOIS, *De la maniere de negocier avec les Souverains* (1716) in Satow (1968).
DE MARTENS, CHARLES, *Guide Diplomatique* (1821) in SATOW (1968).
DE WICQUEFORT, ABRAHAM, *L'Ambassadeur et ses Fonctions* (1679) in Satow (1968).
DER DERIAN, JAMES, *On Diplomacy* (Oxford: Blackwell, 1987).
DLUGOLECKI, ANDREW, '*Environmental Risk Management: What Influences Business to Make Environmentally Responsible Decisions*', Public lecture delivered at the London School of Economics, 17 March 1998.
DONALDSON, JOHN (ed), *Business Ethics: A European Casebook* (London: Academic Press, 1992).
DONALDSON, THOMAS, *The Ethics of International Business* (New York: Oxford University Press, 1989).
DONALDSON, THOMAS and THOMAS W. DUNFEE, *Ethics in Business and Economics* (Aldershot: Ashgate, 1997).
EASTERBROOK, GREG, *A Moment in the Earth: The Coming Age of Environmental Optimism* (New York: Viking, 1995).
Economist, The, 29 July 1995.
Economist, The, 19 August 1995, pp. 33–4; 65–6.
Economist, The, Survey – Technology Industry, 29 March 1997.
Economist, The, 'A Survey of Development and the Environment', 21 March 1998, p. 4.
EDWARDS, MICHAEL, ELLEN LEOPOLD and MICHAEL GEDDES, *King's Cross Railway Lands Redevelopment: Second Report* (London: University College, Bartlett School of Architecture and Planning, 1990).
Electric Vehicle Information Network, 'EV Facts & Myths', 1997 website of the Electric Vehicle Association of Greater Washington, DC (http://www.radix.net/ (1/27/99) 3.
Environmental Resources Ltd and MONTAGU EVANS *et al.*, *London Regeneration Consortium Proposals, King's Cross Railway Lands* (London, 1989).
Financial Times, 'Energy Efficiency Survey', 28 November 1994, p. II.
Financial Times, London, 4 April 1995.
Financial Times, London, 21 June 1995.
FORESTER, T., *High-Tech Society* (Oxford: Blackwell, 1987).
GILDER, GEORGE, *Wealth and Poverty* (New York: Basic Books, 1981).
Global climate coalition, 'The Global Climate Coalition Membership Backgrounder', 21 May 1998, p. 2.
GOLDENBERG, S., 'Silicon City Creaks Under the Strain of Its Success', *The Guardian* (24) 10.
GRAHAM, MARY, 'The Quiet Drug Revolution', *The Atlantic*, Issue no. 267 (Jan. 1991).

Green Globe Yearbook (The Fritjof Nansen Institute, 1988).
Greenpeace, 'Greenpeace Stories', Interview by MICHAEL FRIEDRICH of JIM BOHLEN and DOROTHY METCALF, Greenpeace web site: www.greenpeace.org
Greenpeace International, *1997 Annual Report*, 'Greenpeace International pro forma summary financial statement, years ended 31 December 1997 and 1996, including KMPG Accountants, Amsterdam, Auditor's Report, 15 July 1998. (Greenpeace: www.greenpeace.org/report97/finance2., downloaded 9 November 1998.)
Greenpeace publication 'Overview', March 1997, L3, P28.
Greenpeace web site and Kate Kelland, 'Greenpeace Co-founder Revisits Early Campaign Days', 29 September 1995, Reuters News Service, Amsterdam.
Greenpeace, 'Auditor's Report', *Financial Statements, Greenpeace International pro forma summary statements for the years ended 31 December 1997 and 1996* (www.greenpeace.org/report97/finance2) 1.
Greenpeace, *25 Years of Environmental Activism*, Greenpeace website document.
Greenpeace, *The Turning of the Spar*, published by Greenpeace, London, 1998 (ISBN871532 99X) 215.
Guardian The, 15 January 1999, pp. 3, 15.
Guardian The, 19 January 1999, p. 12.
Guardian The, 8 February 1999, p. 5.
Guardian The, 23 February 1999, pp. 10, 20.
Guardian The, Eurostat data, 10 February 1999, p. 7.
HALPERN, SHANNA L., *The United Nations Conference on Environment and Development: Process and Documentation* (Academic Council on the United Nations System, Reports and Papers, 1993) no. 2, 19.
HAMEL, GARY and C. K. PRAHALAD, *Competing for the Future* (Boston: Harvard Business School Press, 1996).
HAMMER, M. and J. CHAMPY, *Reengineering the Corporation: A Manifesto for Business Revolution* (New York: HarperCollins, 1993) (London: Nicholas Brealey, 1994) pp. 50–6.
HANDY, C., *The Empty Raincoat* (London: Heinemann, 1992).
HEERE, A.H.L., *History of the Political System of Europe and Its Colonies* (1809) in Satow (1968).
HEINONEN, JORM, 'Small Firms and Environmental Technology: An International Perspective', *GMI Journal*, vol. 8 (1994) pp. 37–47.
HERZ, JOHN H., 'Rise and Demise of the Territorial State', *World Politics*, vol. 9, no. 4 (July 1957).
HILL, NAPOLEON, *Think and Grow Rich* (Los Angeles: Wilshire Book Co., 1935, 1st. edn; UK: Thorsons, 1966 edn).
HOLZGREFE, J. L., 'The Origins of Modern International Relations Theory', *Review of International Studies*, vol. 15, no. 1 (January 1989) no. 11–26.
HUGHES, BARRY B., *Continuity and Change in World Politics: The Clash of Perspectives*, 2nd edn (Englewood Cliffs, New Jersey: Prentice-Hall, 1994).
HYMAS, CHARLES and STEVEN HAYNES, 'Greenpeace's Millions Hidden Around the World', *Sunday Times*, London, 22 October 1995.

International Chamber of Commerce, *The Business Charter in Action: Principles for Environmental Management, Summary Report* (ICC Seminar, May 1994).

JACKSON, JENNIFER, *An Introduction to Business Ethics* (Oxford: Blackwell, 1996).

JACKSON, KEVIN T., 'Globalizing Corporate Ethics Programs: Perils and Prospects', in Marilynn Fleckenstein, Mary Maury, Laura Pincus and Pat Primeaux (eds), *From the Universities to the Marketplace: The Business Ethics Journey*, The Second Annual Vincentan Conference (Boston: Kluwer Academic Publishers, 1997) pp. 17–25.

JACKSON, ROBERT H. and CARL G. ROSBERG, 'Why Africa's Weak States Persist: The Empirical and Juridical Statehood', *World Politics*, vol. 35, no. 1 (October 1982).

JACOBS, JANE, *Life and Death of Great American Cities* (New York: Doubleday, 1963).

JACQUES, MARTIN, 'Where the Future is Outrageous', *The Guardian*, 22 March 1997, p. B5.

JAMES, ALAN, 'Diplomacy', *Review of International Studies*, vol. 19, no. 1 (January 1993) p. 95.

JAMES, P., 'Quality and the Environment: From Total Quality Management to Sustainable Quality Management', *Greener Management International, Journal of Corporate Environmental Strategy and Practice (GMI)*, Issue 6 (April 1994) pp. 62–71.

JOHNSON, GERRY and KEVAN SCHOLES, *Exploring Corporate Strategy* (Hempel Hempstead: Prentice Hall Europe, 1993).

JONES, MURRAY, 'The Role of Stakeholder Participation: Linkages to Stakeholder Impact Assessment and Social Capital in Camisea, Peru', *Greener Management International*, Issue 19, (Autumn 1997) pp. 88–97.

KEENS-SOPER, H.M.A. and KARL W. SCHWEIZER (eds), *The Art of Diplomacy* (Leicester University, 1983).

KEENS-SOPER, MAURICE, 'Abraham de Wicquefort and Diplomatic Theory', University of Leicester Diplomatic Studies Programme, No. 14 (February 1996).

KEOHANE, ROBERT O. and ELINOR OSTROM, *Local Commons and Global Interdependence: Heterogeneity and Cooperation in Two Dimensions* (Boston: Center for International Affairs, Harvard University, Sage, 1995).

KITSON, ALAN and ROBERT CAMPBELL, *The Ethical Organisation: Ethical Theory and Corporate Behaviour* (Basingstoke: Macmillan Business, 1996).

KOTLER, PHILIP, DONALD H. HAIDER and IRVING REIN, *Marketing Places: Attracting Investment, Industry and Tourism to Cities, States and Nations* (New York: Free Press, 1993).

LANDES, D., *The Wealth and Poverty of Nations* (London: Heinemann, 1997).

LARSEN, S. and J. ROGERS, *Silicon Valley Fever* (New York: Doubleday, 1984).

LEDGERWOOD, GRANT, *Urban Innovation: London's Docklands, 1968–1984* (Aldershot: Gower, 1985).

LEDGERWOOD, G. (ed.), *Greening the Boardroom* (Sheffield, UK: Greenleaf, 1997).

LEDGERWOOD, G., 'New Corporate Governance Paradigms for Transnational Enterprises – Shell and Strategic Futures for Big Oil', *Corporate Governance – An International Review*, vol. 6, no. 4 (October 1998) pp. 269–77.

LEDGERWOOD, GRANT and ARLENE IDOL BROADHURST, 'Creating Technology-Based Enterprise Televillages: Post-Modern Regional Development Theory', *Cities, The International Journal of Urban Policy and Planning*, vol. 16, no. 1 (February 1999) pp. 43–50.

LEDGERWOOD, G. and E. STREET, 'Corporate Strategy and Environmental Sustainability: Establishing an Environmental Market Ethos to Gain Competitive Advantage', *GMI Journal of Corporate Environmental Strategy and Practice*, vol. 1, no. 3 (July 1993) pp. 41–50.

LEDGERWOOD, G., E. STREET and R. THERIVEL, *The Environmental Audit and Business Strategy: A Total Quality Approach* (London: Pitman Financial Times, 1992).

LEDGERWOOD, GRANT, E. STREET and R. THERIVEL, *Implementing and Environmental Audit – How to Gain Competitive Advantage Through Quality and Environmental Management* (Burr Ridge, Illinois: Irwin Professional Publishing, 1994).

LEDGERWOOD, GRANT, IAN HAYWOOD, NORMAN SHEPPARD and ELIZABETH STREET, 'Environment Assessment as a Real Estate Management Protocol', *Cities*, vol. 11, no. 3 (1994), pp. 181–94.

LEGAULT, ALBERT, CRAIG N. MURPHY and W. B. OFUATEY-KODJE, *The State of the United Nations, 1992*, The Academic Council on the United Nations Systems, Reports and Papers, no. 3 (1992) pp. 1–75.

LICHFIELD, D. and LICHFIELD, N., *Community Impact Evaluation: Principles and Practice* (London: University College Press, 1988).

LILLINGTON, KARLIN, 'Walk on the Wired Side', *The Guardian Online*, Thursday, 14 January 1999, p. 2.

MADDOX, BRONWEN, 'Rubber Suits Turn the Tide for Greenpeace', *Financial Times*, London, 21 June 1995.

MALTZ, MAXWELL, *PsychoCybernetics: A New Technique for Using Your Subconscious Power* (North Hollywood, CA: Wilshire Publishing, 1960).

MANN, HARRIET, 'The Pooh Stakehold', Cambridge MA, unpublished thesis, 1972.

MEADOWS, DONELLA H., DENNIS L. MEADOWS, JORGEN RANDERS and WILLIAM W. BEHRENS III, *The Limits of Growth: A Report for the Club of Rome's Project on the Predicament of Mankind* (Cambridge, MA: MIT Press, 1972).

MEADOWS, D. *et al.*, *Beyond the Limits of Growth* (New York: Chelsea Park, 1992).

MEIKLE, JAMES, 'Greenpeace Dumps Two More of Its Warriors', *The Guardian*, London, 27 January 1996.

MELCHETT, PETER, 'Greenpeace Not Principally a Charity', Letters, *Financial Times*, 21 September 1995.

MINUS, PAUL (ed.), *The Ethics of Business in a Global Economy* (Boston and London: Kluwer Academic Publishers, 1993).

MORAN, THEODORE H. (ed.), *Multinational Corporations* (Lexington, Mass: Lexington Books, 1985).

MORELL, VIRGINIA, 'In Search of Solutions', *National Geographic*, vol. 195, no. 2 (February 1999), p. 84.

MORPHET, SALLY, 'NGOs and the Environment', in Peter Willetts (1996).

MOSER, TITUS and DAMIAN MILLER, 'MNCs' Impacts on the Environment and Communities in the Developing World: A Synthesis of the Contemporary Debate', *Greener Management International*, vol. 19 (Autumn 1997) pp. 40–51.

Newsweek, 22 February 1999, p. 6.

NICOLSON, HAROLD, *Diplomacy* (London: Oxford University Press, 1939, 1950).

NUMELIN, RAGNER, *The Beginnings of Diplomacy: A Sociological Study of Inter-Tribal and International Relations* (Oxford: Oxford University Press, 1950), p. 14.

OATES, WALLACE E. (ed.), *The Economics of Environmental Regulation*, Economists of the Twentieth Century Series (Cheltenham: Edward Elgar, 1996), pp. 675–740.

OSBORNE, D. and P. GAEBLER, *Re-inventing Government* (New York: Addison-Wesley, 1992).

PARKES, MICHAEL, *King's Cross Railwaylands: People's Brief* (London: King's Cross Railwaylands Group, KXT/KXRLG, December 1990).

PEARSON, DAVID, *The Natural House Book* (New York: Simon & Schuster, 1989).

PEARSON, DAVID, *Earth to Spirit: The Search for Natural Architecture* (London: Gaia Books Ltd, 1994) (USA edition, 1995).

PEATTIE, K., *Environmental Marketing Management* (London: Pitman, 1995).

PETRELLA, RICARDO, *Limits to Competition: Towards a New Social Contract* (Cambridge, Mass: MIT Press, 1995).

PORTER, MICHAEL E., *The Comparative Advantage of Nations* (London: Collier Macmillan, 1989).

PREST, M. and S. CAULKIN, 'Where There's Muck: Tarmac', *The Observer*, Business, 26 May 1995, p. 4.

Prince of Wales Business Leaders Forum (www.oneworld.org/pwblf/intor/social.htm) 2 February 1999, p. 1.

RAMO, JOSHUA COOPER, 'The Three Marketeers', *Time*, 15 February 1999, p. 25.

REGISTER, RICHARD, *Ecocity Berkeley: Building Cities for a Healthy Future* (Berkeley, Calif.: North Atlantic Books, 1987).

RICE, ROBERT, 'Greenpeace Steps Up Campaigns', *Financial Times*, London, 11 September 1995.

RIDLEY, MATT, 'Greenpeace – A Cause Driven by Cashflow', *Sunday Telegraph*, London, 10 Sept. 1995.

ROBBINS, ANTHONY, *Unlimited Power* (New York: Fawcett Columbine, 1986).

ROBBINS, NICK and SARAH ROBERTS, 'Trade Opportunities for Developing Country Producers', *Greener Management International*, vol. 19 (Autumn 1997), pp. 53–66.

ROSENAU, JAMES N., *The United Nations in a Turbulent World* (Boulder, Co: Lynne Reinner, 1992).

ROSENAU, JAMES, *Along the Domestic–Foreign Frontier* (Cambridge: Cambridge University Press, 1997).

RUSHKOFF, DOUG, *Cyberia: Life in the Trenches of Hyperspace* (San Francisco: Harper, 1993) p. 21.

RUSSELL, TREVOR, 'World Review', *Greener Management International*, vol. 19 (Autumn 1997), pp. 9–10.

SARAGOSA, MANUELA, 'Freeport Mine Challenged on Environment', *Financial Times*, 20 April 1996, p. 3.

SATOW, RT. HON. SIR ERNEST, *A Guide to Diplomatic Practice* (London: Longmans, Green & Co., 1932).

SCALLON, M. and M. STEN, 'Environmental Positioning for the Future', *GMI Journal of Corporate Environmental Strategy & Practice*, vol. 13 (January 1996), pp. 49–65.

SCHMIDHEINY, STEPHAN and the Business Council for Sustainable Development, *Changing Course: a Global Business Perspective an Development and the Environment* (Cambridge, Mass.: MIT, 1992).

SCHMIDHEINY, STEPHAN, RODNEY CHASE and LIVIO DESIMONE, *Signals of Change: Business Progress Towards Sustainable Development* (Geneva: WBCSD Publications, undated, 1998?) p. 4.

SCHWARTZ, D. J., *The Magic of Thinking Big* (North Hollywood, California: Wilshire Book Co., 1959).

SERVAN-SCHREIBER, JEAN-JACQUES, *Le défi américain* (London: Hamish Hamilton, 1968).

Shell Group, *Health, Safety and Environment Report, 1997* (The Hague: Royal Dutch/Shell Group of Companies, 1997).

Shell International, *Shell and the Environment* (London: Group Public Affairs, Shell International Petroleum Co., 1995).

Shell UK, *Environmental Report 1995* (London: Shell UK Limited, 1995).

Shell UK Ltd, *Report to Society* (May 1998).

SHELTON, R. J. and J. SHOPLEY, 'Beyond the Green Wall: Rethinking the Environment for Business Advantage', in G. LEDGERWOOD (ed.), *Greening the Boardroom* (1997), pp. 110–28.

SHRIVASTAYA, PAUL, 'Greening of Business', in D. SMITH (ed.), *Business and the Environment: The Implications of the New Environmentalism* (London: Paul Chapman Publishing, 1993).

SHULMAN, JENNIFER B., 'The North American Free Trade Agreement (NAFTA): What It Is and How It Came To Be', in MADELEINE O. HOSLI and ARILD SAETHE (eds), *Free Trade Agreements and Customs Unions: Experiences, Challenges and Constraints* (Brussels: European Commission; and Maastricht, the Netherlands: The European Institute of Public Administration, 1997), pp. 216–18.

SIMON, HERBERT, *The Art of Decision* (New York: Macmillan, 1957).

SMITH, DUVALL (ed.), *Business and the Environment: Implications of the New Environmentalism* (London: Paul Chapman Publishing, 1993).

SOLERI, PAOLO, *Arcology: The City in the Image of Man* (Cambridge, Mass.: MIT Press, 1969).

SOLOMON, ROBERT C., *Ethics and Excellence: Co-operation and Integrity in Business* (New York: Oxford University Press, 1993).

SOLOMON, R. C., *The New World of Business: Ethics and Free Enterprise in the Global 1990s* (UK: Littlefield Adams Quality Paperbacks, 1994).

SOREL, ALBERT, *L'Europe et la Revolution française* (1885) in Satow (1968).

SORSA, P., 'Competitiveness and Environmental Standards', Working Paper No. 1249 (Washington, D.C.: The World Bank, International Economics Department, 1994).

SPERO, JOAN EDELMAN, *The Politics of International Economic Relations*, 4th edn (New York: St. Martin's Press, 1990).

STAIRS, KEVIN and PETER TAYLOR, 'NGOs and Legal Protection of the Oceans', in Andrew Hurrell and Benedict Kingsbury (eds), *The International Politics of the Environment* (Oxford: Clarendon Press, 1992), p. 113.

STARR, GARY, 'The Shocking Truth About Electric Cars', Solarelectric (http://metalab.unc.edu/) pp. 1–3.

STERNBERG, ELAINE, *Just Business: Business Ethics in Action* (London: Little Brown & Co., 1994).

STUART, J., 'World Review – April 1995', *Greener Management International, Journal of Corporate Environmental Strategy and Practice* (April 1995), pp. 6–33.

SUSSKIND, LAWRENCE E., *Environmental Diplomacy: Negotiating More Effective Global Agreements* (New York and London: Oxford University Press, 1994).

TEN BRINK, PATRICK and R. HAINES, 'World Environmental Industries: Market Drivers, Developments and Implications for Corporate Strategies', *GMI Journal*, vol. 8 (1994) pp. 26–36.

TENENBAUM, D., 'The Greening of Costa Rica', *Technology Review* (October 1995) pp. 42–53.

TOYDER, GEORGE, *The Just Enterprise: A Blueprint for the Responsible Company* (London: Adamantine Press, 1993).

UN Framework Conventon on Climate Change, Kyoto Protocol, art. 13, no. 8, 15.

US Department of State, 'Environmental Diplomacy: The Environment and US Foreign Policy', *Annual Report on the Environment and Foreign Policy*, 1997.

VAN DEN BERGH, JEROEN C. J. M., *Ecological Economics and Sustainable Development: Theory, Methods and Application* (Cheltenham: Edward Elgar, 1996).

VELASQUEZ, MANUEL, *Business Ethics: Concepts and Cases*, 4th edn (London: Prentice Hall International, 1998).

VERNON, RAYMOND, *Sovereignty at Bay* (New York: Basic Books, 1971).

VIDAL, J., 'Crisis, What crisis?', *The Guardian*, 24 July 1996, pp. 4–5.

WEBB, C., 'Dial the Kettle', *Sunday Times*, 5 January 1999, p. 4.

WEBBER, M., 'The Non-Place Urban Realm and the New Metropolitan Structure', in Donald Foley (ed.), *Explorations in Urban Structure* (Philadelphia: University of Pennsylvania Press, 1967), pp. 79–153.

WHIPPLE, THOMAS W. and DOMINIC F. SWORDS, *Business Ethics Judgements: A Cross-Cultural Comparison Between the USA and the UK*

(Henley-on-Thames, Oxon: The Management College, Working Paper Series, 1991).

WHITE, L. C., *International Non-Governmental Organizations* (New Brunswick: Rutgers University Press, 1951).

WHITE, R. F. and S. FRALEY, 'Imperfect Competition, Price Fairness and the Pharmaceutical Industry', *On Line Journal of Ethics*, 1997 (DePaul University, Institute of Business and Professional Ethics, September 1998): http://www.depaul.edu/ethics/.

WIEMAN, CLARK, 'Downsizing Infrastructure', *Technology Review* (May 1996), pp. 48–55.

WIENER, MARTIN J., *English Culture and the Decline of the Industrial Spirit (1850–1980)* (Cambridge University Press, 1981).

WILLETTS, PETER (ed.), *The Influence of Non-Governmental Organisations in the U.N. System* (London: The David Davies Memorial Institute of International Studies, and Hurst & Co., 1996), p. 56.

WILLUMS, JAN-OLAF and ULRICH GOLUKE, *From Ideas to Action: Business and Sustainable Development: The ICC Report on the Greening of Enterprise* (Oslo: ICC Publishing and Ad Ad Notam Gyldendal, 1992).

WINTER, L. and G. LEDGERWOOD, 'Motivation and Compliance in Environmental Performance for Small and Medium-sized Businesses', *GMI Journal, Corporate Environmental Strategy and Practice*, vol. 7 (July 1994), pp. 62–73.

World Business Council for Sustainable Development, *Eco-Efficient Leadership for Improved Economic and Environmental Performance*, 1966.

World Business Council for Sustainable Development, *Annual Review 1996* (UK: Beacon Press, 1997), p. 1.

World Business Council for Sustainable Development newsletter, *Sustain*, Issue 3 (April 1997) pp. 1–3.

World Business Council for Sustainable Development and The World Conservation Union (IUCN), *Business and Biodiversity: A Guide for the Private Sector*, June 1997.

www.depaul.edu/ethics/

Www.simonjones.org.uk

www.nir.org/NRET/nret.htm

YOUNG, JAMES HARVEY, *The Toadstool Millionaires: A Social History of Patent Medicines in America Before Federal Regulations* (Princeton New Jersey: Princeton University Press, 1972).

Index